**LABOR AND EMPLOYMENT
RELATIONS ASSOCIATION SERIES**

# The Climate–Labor Movement: Lessons Learned and the Promise of an Equitable and Diverse Clean Energy Economy

Edited by

Richard A. Benton
Lara Skinner

Cover design: Mary Cha
Cover photo: Freepik

First Edition
ISBN 978-0-913447-31-4
Price: $39.95

## LABOR AND EMPLOYMENT RELATIONS ASSOCIATION SERIES

LERA *Proceedings of the Annual Meeting* (published online annually, in the fall)

LERA *Annual Research Volume* (published annually, in the summer/fall)

LERA Online Membership Directory (updated daily, member/subscriber access only)

LERA *Labor and Employment Law News* (published online each quarter)

LERA *Perspectives on Work* (published annually, in the fall)

Information regarding membership, subscriptions, meetings, publications, and general affairs of LERA can be found at the Association website at www.leraweb.org. Members can make changes to their member records, including contact information, affiliations, and preferences, by accessing the online directory at the website or by contacting the LERA national office.

LABOR AND EMPLOYMENT RELATIONS ASSOCIATION
University of Illinois Urbana-Champaign
School of Labor and Employment Relations
121 Labor and Employment Relations Building
504 East Armory Ave., MC-504
Champaign, IL 61820
Telephone: 217/333-0072   Fax: 217/265-5130
Website: www.leraweb.org
E-mail: LERAoffice@illinois.edu

# Contents

**Introduction** ........................................................................................... 1
*Richard A. Benton*

**Chapter 1:** Green Jobs, Climate Jobs, and Competing Visions for a New
Energy Economy .................................................................................... 9
*Brendan Davidson and Dimitris Stevis*

**Chapter 2:** Working Conditions in the U.S. Solar Industry: Findings and
Learnings from Studies in New York and Texas ................................. 37
*Jillian Morley and Avalon Hoek Spaans*

**Chapter 3:** Organizing a Worker- and Community-Centered Transition:
The Contra Costa Refinery Transition Partnership as Case Study ............. 59
*Virginia Parks and Jessie HF Hammerling*

**Chapter 4:** Fracking or No Fracking? How a Green Transition Can
Work for Workers ................................................................................ 75
*Robert Pollin and Jeannette Wicks-Lim*

**Chapter 5:** From Here to There—Advancing a Just Transition for All ......... 103
*Todd E. Vachon*

**Chapter 6:** Climate Jobs and Manufacturing: Green Industrial Policy Must
Mean Good Jobs ................................................................................ 119
*Mike Williams*

**Chapter 7:** Stronger Together: The Role of Sectoral Bargaining in
Advancing a Just Transition for Autoworkers ................................... 137
*Hunter Moskowitz and J. Mijin Cha*

**Chapter 8:** Industrial Environmental Policy: Markets, Labor, and
the Rhode Island Experiment ............................................................ 153
*Patrick Crowley*

**Chapter 9:** Building a Diverse, Equitable, and Unionized Clean Energy
Workforce: Best Practices and Lessons Learned ............................... 167
*Zach Cunningham and Melissa Shetler*

**About the Contributors** ........................................................................ 191

**LERA Executive Board Members 2025–26** ........................................ 197

# Building a Just and Durable Clean Energy Economy: Context, Themes, and Tensions Across the Volume

RICHARD A. BENTON

*University of Illinois Urbana-Champaign*

## THE STAKES OF THE TRANSITION

The accelerating climate crisis and deepening economic inequality have converged to make the question of how we transition to a clean energy economy one of the most urgent challenges of our time. The stakes of the green transition are profound and multifaceted. On one hand, the scientific consensus is clear: decarbonization must proceed at an unprecedented pace to avert catastrophic warming. On the other, the social and economic structures that underpin our current energy system are deeply unequal, with the burdens of pollution, job loss, and economic insecurity falling disproportionately on working-class communities, communities of color, and regions historically dependent on fossil fuel industries. The transition to a low-carbon economy, if left to market forces alone, risks reproducing or even deepening these inequalities.

The stakes of the energy transition extend far beyond the technical challenge of reducing greenhouse gas emissions. At their core, they involve fundamental questions about justice, democracy, and the future of work. The transition is not simply about swapping out one set of technologies for another; it is about reimagining the social contract that binds together workers, communities, and the state.

Historically, the costs of environmental degradation and economic restructuring have not been borne equally. Communities of color and working-class neighborhoods have long been situated near polluting industries, power plants, and hazardous waste sites, suffering higher rates of illness and economic dislocation. As fossil fuel industries contract, these same communities are often the first to experience job losses and the last to benefit from new investments in clean energy. Without deliberate intervention, the energy transition could replicate these patterns in both rural and urban U.S. communities.

Moreover, while the clean energy sector is often celebrated for its job creation potential, research—including studies in this volume—shows that many new "green jobs" are characterized by precarity, low wages, and limited benefits, especially in nonunionized segments like solar installation and some energy services. The rapid growth of these sectors, if not accompanied by strong labor standards and pathways to unionization, risks entrenching a two-tier labor market, one in which a minority of workers enjoy stable, high-quality employment, while many others face insecurity and exploitation.

The geography of the energy transition further complicates these challenges. Regions historically dependent on coal, oil, and gas—such as Appalachia, the Gulf Coast, and parts of the Midwest—face acute risks of economic decline, population loss, and fiscal crisis as fossil fuel industries contract. At the same time, the benefits of clean energy investment are often concentrated in regions with existing infrastructure, skilled workforces, and political support for renewables. This unevenness raises urgent questions about regional development, economic diversification, and the role of public policy in managing transition.

If the transition to a low-carbon economy is left primarily to market forces, it is likely to reproduce the same inequalities that have defined the fossil fuel era. Market-driven approaches tend to prioritize cost minimization and short-term returns, often at the expense of job quality, community stability, and long-term social investment. As several authors in this volume argue, a just transition requires proactive public policy: investments in workforce development, social safety nets, community-led planning, and democratic governance of energy systems.

At a time when the political environment is uncertain, the legitimacy and durability of the energy transition ultimately depend on its perceived fairness. When workers and communities feel excluded or left behind, they may become powerful opponents of climate policy, fueling political backlash and undermining the prospects for sustained decarbonization. Conversely, when the transition is seen as an opportunity to build a more equitable and democratic society, it can become a source of hope and mobilization.

This volume brings together a diverse set of scholars, practitioners, and labor leaders to examine how the transition can be not only rapid and effective but also just, inclusive, and democratically governed.

The chapters in this collection reflect a growing field of interdisciplinary inquiry and practice that centers the role of workers, labor institutions, and industrial policy in shaping the contours of the green transition. They offer grounded case studies, theoretical interventions, and policy frameworks that grapple with the complexities of decarbonization in a deeply unequal society. Together, they illuminate the stakes of the transition—not just for emissions targets, but for the future of work, democracy, and social equity.

This volume comes together at a moment of both opportunity and uncertainty. Recent years have seen historic investments in clean energy and infrastructure—such as the Inflation Reduction Act and the Infrastructure Investment and Jobs Act—alongside new commitments to equity and job quality. Yet, as several authors in this collection note, these gains are fragile and contested, especially in the wake of shifting political winds and the rollback of federal climate and labor protections. The future of climate policy, job quality standards, and equity initiatives is now uncertain, making the role of state and local actors, labor unions, and community organizations more critical than ever.

In bringing together empirical research, policy analysis, and on-the-ground perspectives, this volume aims to highlight the stakes of the transition—not just for emissions targets but for the future of work, democracy, and social equity. It offers both cautionary tales and hopeful models, underscoring that a just transition is not inevitable but must be actively constructed through struggle, innovation, and solidarity.

## KEY THEMES ACROSS THE VOLUME

Several themes cut across the chapters. Taken together, these recurring threads demonstrate the broader significance of the volume's contributions. By highlighting patterns, tensions, and shared insights that surface in diverse contexts, this section emphasizes underlying dynamics that shape the challenges and possibilities of a just energy transition. The following discussion draws out these overarching themes, providing a framework for interpreting the chapters as part of a larger, interconnected conversation.

### The Centrality of Organized Labor

Across the volume, nearly every chapter foregrounds the role of organized labor in shaping the clean energy transition and centering workers' interests in industrial policy, planning, and economic development. Rather than treating unions and worker organizations as peripheral stakeholders, the chapter authors collectively position labor as a central participant—shaping the terms of decarbonization as well as the broader contours of economic and social transformation. For instance, Williams's critique of green industrial policy highlights the risks of relying on market incentives and voluntary standards, arguing that without robust labor requirements, as advocated by organized labor, the promise of new manufacturing and clean energy jobs may simply reproduce the precarity and inequity of the past. Similarly, Cunningham and Shetler's analysis of apprenticeship readiness programs demonstrates how union-led training and hiring pathways are essential for building a skilled, diverse, and resilient workforce capable of meeting the demands of large-scale infrastructure and climate projects while expanding access to high-quality jobs. Moskowitz and Cha make the case for sectoral bargaining in the electric vehicle industry, contending that only through collective action and industry-wide standards can workers secure fair wages, safe conditions, and a meaningful voice in shaping the future of work.

Taken together, these chapters underscore that organized labor is not merely a constituency to be consulted but a dynamic force capable of negotiating, implementing, and sustaining the policies and practices that will define the green transition. Whether through collective bargaining, policy advocacy, or the design of workforce development systems, unions and worker organizations are central to any effort to ensure that decarbonization is not achieved at the expense of job quality, equity, or democratic participation. The contributors thus insist that any serious strategy for climate action must place labor at the center—not only as a matter of justice but as a practical necessity for building durable, inclusive, and broadly supported pathways to a sustainable future.

### Place-Based Strategies

Place-based strategies are another central theme across the volume, underscoring the necessity of tailoring just transition efforts to the unique characteristics of specific regions and communities. By grounding strategies in the specific social, economic, and political contexts of particular regions, place-based efforts can more effectively address the realities workers and communities face. These approaches recognize that labor markets, industrial legacies, and community priorities vary widely across the United States, and that successful transition policies must be responsive to these differences. Place-based strategies also foster stronger local buy-in, enable more meaningful participation from affected stakeholders, and can leverage existing networks and institutions to build durable solutions. Especially in an era of federal retrenchment, state and local initiatives have become critical laboratories for innovation, demonstrating how tailored policies can drive both decarbonization and equity.

The chapters in this volume illustrate the power and necessity of place-based strategies through a range of concrete examples. For instance, Pollin and Wicks-Lim's modeling of a fracking phaseout in Pennsylvania shows how region-specific economic analysis can inform targeted support for displaced workers. Similarly, Vachon's discussion of local hiring requirements highlights how labor policies can be designed to benefit communities most immediately impacted by the transition. Crowley's examination of Rhode Island's industrial policy demonstrates the potential for state and local governments to lead on climate and labor issues, even when federal action stalls. Morley and Hoek Spaans's comparative studies of the solar workforce in New York and Texas reveal how working conditions and labor market structures differ across states, underscoring the need for regionally tailored interventions. Parks and Hammerling's case study of Contra Costa County emphasizes the importance of local economic planning and community-driven transition strategies, while Cunningham and Shetler's analysis of apprenticeship readiness programs highlights the value of local partnerships and recruitment attuned to regional labor market dynamics. Collectively, these authors make clear that effective just transition strategies must be rooted in the particular histories, institutions, and aspirations of the places in which they unfold.

## The Role of Worker Voice and Democratic Participation in Policy Planning

Energy transition strategies are most effective when they center worker voice and democratic participation in policy planning. By involving workers, unions, and community members directly in the design and implementation of transition policies, these approaches ensure that solutions are grounded in lived experience and local knowledge and represent the material interests of those they are intended to serve. Democratic participation increases the legitimacy and effectiveness of transition efforts and helps incorporate the diverse needs and aspirations of different communities. When workers and residents have a meaningful say in shaping the future of their local economies, policies are more likely to address real concerns—such as job quality, economic security, and environmental health—rather than imposing one-size-fits-all solutions imposed from above. This participatory approach also builds trust, strengthens local institutions, and fosters a sense of shared ownership over the transition process, making outcomes more durable and equitable.

The chapters in this volume contain concrete examples of how worker voice and democratic participation can shape transition strategies. Rather than treating workers as passive recipients of change, several contributors argue that meaningful participation by workers and their organizations is essential for designing effective and equitable policies. For instance, Davidson and Stevis trace the historical relationship between labor and environmental movements, highlighting how the inclusion—or exclusion—of worker perspectives has shaped both workplace transformation and broader policy outcomes. Parks and Hammerling's case study of Contra Costa County demonstrates how codetermination, with unions and community groups sharing leadership, can foster more responsive and legitimate transition planning. Vachon emphasizes the value of worker representation on decision-making boards, ensuring that those most affected by the energy transition have a direct role in shaping its direction. Williams discusses the role of worker organizing and the integration of worker input into community benefit plans, while Crowley details how union participation in Rhode Island's legislative and implementation processes has influenced the design and rollout of climate policy. Cunningham and Shetler further illustrate how worker voice can be embedded in the design and evaluation of workforce development programs, including through participation on workforce boards. Collectively, these chapters show that centering worker experience and democratic participation is not only a matter of justice but also a practical necessity for crafting policies that are grounded, effective, and broadly supported.

## TENSIONS AND COMPLEMENTARITIES

Several productive tensions animate the volume. Across the chapters, authors grapple with competing priorities, trade-offs, and differing perspectives on how to achieve a just energy transition. At the same time, the collection highlights areas of complementarity, where diverse approaches and strategies reinforce one another or open up new possibilities for collaboration. This section explores how these tensions and complementarities shape the debates, policy choices, and practical pathways that emerge throughout the volume.

### Scale and Specificity

The chapters in this volume reveal a tension between the need for scalable, broadly applicable frameworks and the realities of regional specificity. Some contributors, such as Williams, emphasize national-level policy recommendations that aim to address the scale of the climate crisis and the scope of economic transformation required. Similarly, Moskowitz and Cha offer sectoral frameworks that cut across geographic boundaries, focusing on the auto and EV industries as a whole. These approaches are valuable for identifying common challenges, setting ambitious targets, and advocating for federal or industry-wide standards that can drive large-scale change.

At the same time, other chapters—such as those by Parks and Hammerling or Crowley—demonstrate the importance of local and regional policy. These contributions show how the dynamics of transition play out differently depending on local labor markets, political cultures, and community histories. Place-based approaches emphasize the need for tailored solutions that respond to the unique assets and vulnerabilities of specific regions, and it often uncovers challenges or opportunities that national models might overlook.

This tension between scale and specificity is not easily resolved. National and sectoral frameworks can provide coherence, mobilize resources, and set minimum standards, but they risk imposing uniform solutions that may not fit local contexts. Conversely, highly localized strategies can foster innovation and community buy-in, but they may struggle to achieve the scale necessary to meet climate goals. The chapters in this volume suggest that effective just transition policy will require ongoing negotiation between these levels—drawing on the strengths of both scalable models and place-based approaches, and remaining attentive to the ways they interact, overlap, and sometimes conflict.

## Technocratic and Democratic Approaches

The chapters also highlight a productive tension between technocratic and democratic approaches to the energy transition. On one hand, several authors emphasize the importance of rigorous policy design, economic modeling, and technical expertise. For example, Pollin and Wicks-Lim draw on quantitative analysis and policy frameworks to map out pathways for decarbonization, job creation, and industrial transformation at scale. Similarly, Williams focuses on policy design incorporating structured incentives, regulatory frameworks, and measurable standards in shaping large-scale coordinated responses. These technocratic approaches are essential for setting ambitious targets, allocating resources efficiently, and ensuring that transition policies are grounded in evidence and capable of meeting the scale of the climate crisis.

On the other hand, several contributors emphasize the necessity of participatory governance, worker voice, and democratic decision making. Parks and Hammerling, for instance, document how community-led planning and coalition building in Contra Costa County have shaped more equitable and durable transition strategies. Moskowitz and Cha explore the role of worker organizing and sectoral bargaining in shaping the future of the auto and EV industries, while Davidson and Stevis draw on the example of the Lucas Plan to illustrate the transformative potential of democratic technological development and workplace governance. These chapters argue that meaningful participation by workers and communities is not only a matter of justice but also a practical requirement for building legitimacy, trust, and long-term support for transition policies.

The volume suggests that both technocratic and democratic approaches are necessary but that their integration requires careful attention. Technocratic expertise can provide the analytical tools and policy levers needed to drive large-scale change, but it risks becoming disconnected from the lived realities and priorities of affected communities if not paired with robust democratic engagement. Conversely, participatory processes can surface local knowledge, build buy-in, and adapt solutions to specific contexts, but they may struggle to deliver the technical rigor or coordination needed for systemic transformation if not supported by expert input. The challenge, as reflected across the chapters, is to create institutional arrangements and policy processes that harness the strengths of both approaches—ensuring that technical solutions are shaped by democratic deliberation, and that participatory governance is informed by sound analysis and planning.

## Labor-Centered and Community-Centered Approaches

Finally, the chapters reveal a tension between labor-centered and community-centered approaches to the energy transition. While all contributors recognize the centrality of workers and the importance of good jobs, some chapters—such as those by Moskowitz and Cha and by Cunningham and Shetler—focus primarily on union pathways, apprenticeship programs, and the role of organized labor in shaping transition outcomes. These chapters emphasize the need for strong labor standards, collective bargaining, and union representation to ensure that new jobs in the green economy are high quality, secure, and accessible to workers who have historically powered the industrial economy.

Other chapters—such as those by Parks and Hammerling or Morley and Hoek Spaans—foreground community and equity concerns, highlighting the experiences and needs of frontline communities, marginalized groups, and workers who may not be reached by traditional union pathways. These contributions stress the importance of inclusive planning, targeted outreach, and policies that address racial, gender, and geographic disparities. They show that community-centered strategies can surface local knowledge, build broader coalitions, and ensure that the benefits of the transition are shared more equitably.

This tension reflects deeper questions about how to build coalitions that do not pit labor and community interests against each other. As Davidson and Stevis discuss in their account of the movement's development, overcoming the "jobs versus environment" frame requires intentional efforts to bridge divides and create shared agendas. The chapters collectively suggest that successful just transition strategies will need to integrate both labor-centered and community-centered approaches—recognizing the power of unions to raise standards and protect workers while also ensuring that the voices and needs of diverse communities are at the heart of transition planning. Building such coalitions is challenging but essential for achieving both equity and effectiveness in the shift to a clean energy economy.

## CONCLUSION: TOWARD A DEMOCRATIC DECARBONIZATION

Taken together, the chapters in this volume show that there is no single blueprint for a just energy transition. Instead, the work ahead will require balancing national frameworks with local realities, integrating technical expertise with democratic participation, and bridging the interests of labor and community. The tensions and trade-offs explored throughout the book are not obstacles to be avoided but essential features of a transition that is both ambitious and fair.

Ultimately, the key takeaway is that the path to a low-carbon future will be shaped by the choices we make about power, participation, and equity. The case studies and analyses here underscore that durable progress depends on centering the voices of those most affected, building strong institutions for collaboration and conflict resolution, and remaining attentive to the specific needs of different places and people. As the landscape continues to shift, the lessons and strategies collected in this volume offer both guidance and inspiration for those committed to making the energy transition work for all.

# CHAPTER 1

# Green Jobs, Climate Jobs, and Competing Visions for a New Energy Economy

BRENDAN DAVIDSON
DIMITRIS STEVIS
*Colorado State University*

## Abstract

The simplicity of "green jobs" belies the contradictory meanings contained within the term, including obscuring what the relation between work and the broader political economy entails. Similar to sustainability or justice, meanings vacillate over time and across political realities and contexts. In this chapter, we offer two explorations. First, we detail the historical development of "green jobs" and "climate jobs" as conceptual constructs in the United States. We close this genealogical account by proposing an analytical scheme that allows for differentiation among green and climate jobs according to their social and ecological, or eco-social, promise. To aid us in this, we modify earlier work by Kate Crowley (1999) by making more explicit the social dimensions of her typology. The chapter ultimately argues against green job creation as a good unto itself. Like other forms of work, green jobs are embedded in a social milieu characterized by the expanding precarity and continuing casualization of work brought on by neoliberalism. Awareness of this social embeddedness requires attentiveness to the histories that have contributed to different visions of what sustainable employment should look like, some of which are more socially and environmentally sensitive than others.

## INTRODUCTION

The passage and implementation of the Inflation Reduction Act (IRA) and Infrastructure Investment and Jobs Act (IIJA) were heralded for a variety of reasons, but their potential employment benefits were at the center. Climate jobs also got their time to shine, with the Civilian Climate Corps suggesting the emergence of a more specific type of green job—those tied to climate. In both cases, the promise of these policies is succinctly captured by the BlueGreen Alliance (2024), which writes, "Americans don't have to choose between creating good jobs and protecting the environment. We can and must do both."

These policies, closely associated with "Bidenomics," did not prevent President Donald Trump's return to the White House. While the reasons for President Trump's re-election are legion, he is symptomatic of a deep dissatisfaction with the current state of the country. Many workers and communities feel left behind, ignored, and resigned to adapting to the vagaries of global capitalism on their own (Brown 2019; Cramer 2016). Given this backdrop, an important question presents itself: what alternative political futures are possible, and what do different incarnations of "green jobs" and "climate jobs" portend?

The apparent simplicity of these terms clouds contestation about them, including among those who are in favor (Bowen, Kuralbayeva, and Tipoe 2018; Stanef-Puică et al. 2022). Like other concepts, ideological preferences produce diverse views on sustainability and the proper role that work (and workers) should play in society. In this respect, different conceptions of "green jobs" or "climate jobs" serve as proxies for the broader political economy and associated pathways through which the energy transition might develop. While many academic studies and gray literature calculate green job requirements for the energy transition, fewer cover both ecological *and* social aspects or the political economy they advance.

To help make these debates a bit more productive, we examine green jobs and climate jobs from two angles. First, we trace the genealogy of the concepts in the United States, with references to events and organizations outside the country as appropriate. Our account makes clear that there is significant variability, both diachronically and synchronically. We then offer a modified typology adapted from Kate Crowley (1999) that highlights both the environmental and the social dimensions, the latter being less evident in her typology. The eco-social typology we propose can, in our view, move us toward a more systematic examination of green job theory and practice by making more apparent what kind of political economy they advance. We also offer our own intervention that transformative work must integrate the social and the ecological across space and time.

## A GENEALOGY OF GREEN JOBS AND CLIMATE JOBS
### Early Examples
Environmental employment programs have existed for decades, stretching at least as far back as President Franklin Roosevelt's Civilian Conservation Corps (CCC) (Maher 2009). A project informed by Roosevelt's own time conserving forest land on his property, the program sent millions of young American men from mostly urban areas to rural places to work on resource conservation projects (Dewey 2019; Maher 2009). The CCC also included opportunities for immigrant youth to become part of the American workforce, building both muscles and American values— including a conservation ethos (Dewey 2019). Like many American programs, however, fewer opportunities were made for African Americans, and those who participated sometimes worked in segregated camps (Dewey 2019). The New Deal also provided employment through the Works Progress Administration and Public Works Administration, which created a great deal of public infrastructure that was considered environmental by the standards of the time.

While the CCC is celebrated as one of the most significant conservation and work programs in the country's history, labor environmentalism in the United States goes back much further (Montrie 2018). Enslaved African Americans in the antebellum South developed an environmental ethos by cultivating land, hunting, and fishing to supplement their meager rations. In so doing, they became more attuned to the landscape (Dewey 2019). In the early 1800s, "mill girls" in New England's textile factories turned to the nearby woodlands to escape the increasingly harsh conditions of industrial capitalism, and their writings helped inspire early labor agitation (Dewey 2019). The famed Wobblies of the Industrial Workers of the World argued for clean and healthy forests and bodies, finding resonance between workers and environmental issues, despite lacking contemporary language consistent with environmental justice (Loomis 2021: 131).

Despite these early examples, modern conceptions of environmentalism or even conservation were somewhat scarce among the broader population (Dewey 2019). Instead, many conservation measures adopted over the course of the 19th century served the interests of the wealthy. Such moves included the criminalization of traditional pursuits like fishing and hunting (Montrie 2018: 55) and the expropriation of indigenous lands through enclosures of a variety of means, including the National Parks (Montrie 2018; Treur 2021).

Even so, groups such as the UAW would institute a Recreation Department in 1939 to provide opportunities for members and their families to bask in the regenerative glow of "wilderness" (Dewey 2019; Montrie 2008). Hunting and fishing surged in the American workforce following World War II, which helped spur efforts to protect game species habitat (Dewey 2019). Groups like the AFL-CIO supported outdoor recreation and the protection of open spaces, in part for anticipated increases in free time from the country's prosperity, increased productivity, and increasing automation (Dewey 1998: 50). Such reasoning led AFL-CIO President George Meany, a par excellence supporter of business-unionism, to support the 1958 National Wilderness Preservation Bill declaring wilderness as a benefit for all American people, not just commercial enterprises or the wealthy elite (Dewey 1998). President Al Hartung of the International Woodworkers of America (IWA) would also testify in favor of the Wilderness Act when Congress debated it in 1960, knowing what would happen to forests if timber companies were granted full control (i.e., they would disappear, along with associated jobs) (Crowley 2024: 73; Loomis 2022).

The 1960s often mark the beginning of "environmentalism" in popular consciousness with the publication of *Silent Spring* by Rachel Carson in 1962 [2002], although Murray Bookchin (1962) also published his book *Our Synthetic Environment* a few months before. Prior to this, however, the 1948 Donora Smog, which killed 20 people in Donora and Webster, Pennsylvania, led local leaders and the United Steelworkers Union to pressure the Public Health Service to investigate the smog (Jacobs et al. 2018). The 1960s also saw the establishment of another Youth Conservation Corps by President Johnson, although the program struggled to succeed in the rural West when young men, often minorities, were dropped into unfamiliar and unfriendly environments (Sanders 2021). However, as the preceding history demonstrates, many working people from a diverse array of backgrounds and experiences moved environmentalism forward, both inside and outside of the workplace.

## Seeking a Response to Jobs Versus the Environment

Labor environmentalism picked up steam after WWII and accelerated during the 1960s and 1970s (Kazis and Grossman 1991 [1982]; Leopold 2007; Miller 1980; Montrie 2018; Stevis 2013). The passage of the Clean Air Act (1963), the National Environmental Protection Act (1970), and the Clean Water Act (1972) marked the new political salience of environmental issues in the United States (Vig and Kraft 2013: 2). Even before the passage of these laws, however, worker activists brought environmentalism into the fold of their unions (Gordon 2004). UAW President Walter Reuther espoused environmental virtues and hosted a 1965 event titled the "United Action for Clean Water Conference" (Dewey 1998: 51). The UAW also brought together unionists, conservationists, and community organizers before the first Earth Day took place in 1970 (Dewey 1998: 52). Indeed, several unions, including the UAW, were early supporters of Earth Day and lent it substantial financial, logistical, and advertising support (Uehlein 2010; Velut 2011). In 1976, the UAW sponsored the first ever conference on environmental justice that brought together unionists and other social and environmental activists (Rector 2018).

Labor environmentalists such as Tony Mazzocchi of the Oil, Chemical and Atomic Workers (OCAW) also worked with Ralph Nader's network to pass the Occupational Safety and Health Act (OSHA) of 1970 (Page and O'Brien 1973). The OCAW executed a 1973 strike and boycott against Shell Oil with support from environmental groups regarding workplace and general environmental conditions (Dewey 2019; Velut 2011). The United Farm Workers would also use the country's newfound environmentalism strategically, playing on consumer concerns about pesticide use to run a successful boycott campaign of the grape industry, helping them win their first contract (Bresnihan and Millner 2023; Dewey 2019; Nicholson 2014). Similarly, the IWA would also leverage environmentalism to promote timber worker protections (Loomis 2021: 137).

The early 1970s were a turning point, however. First, there was a systematic reaction to the mobilizations of the 1960s and early 1970s, reflected in Nixon's Southern Strategy and the infamous Lewis F. Powell memo. Second, and related, there was an unraveling of the post-WWII settlement that had given unions a significant presence in exchange for deradicalization and abandoning strategic decisions to the corporation (Loomis 2018). Accordingly, unions found themselves unable to contain corporate decisions over location, including the U.S. South that had been systematically inviting capital to take advantage of its weak labor politics, in addition to other technological pressures.

Capital flight and deindustrialization enabled corporations to employ "job blackmail" against unions and communities. "Job blackmail" is not limited to the environment but is a general strategy against any countervailing corporate regulation and mobilization. However, it was used to divide unions and environmentalists—with the active participation of conservative business unions and the AFL-CIO—during a time when environmental regulations were questioning a range of corporate (and state) practices (Loomis 2021: 128). This concern is evident in a 1972 presentation by then-UAW President Leonard Woodcock at the Institute of Industrial Relations at the University of California, Berkeley.

In his speech, Woodcock (1972) argued that companies would use "environmental blackmail" to drive a wedge between workers and their communities and the environment, a partial consequence of a 1971 NLRB decision ruling in favor of GM against Local 864 and the UAW stating that unions had no say in "fundamentally managerial issues," such as where company facilities should operate (National Labor Relations Board 1971). Woodcock's speech proved prescient. The confluence of policies hostile to labor, new environmental regulations, and structural features of the world economy (increased globalization and the potential for capital flight, deindustrialization, and automation) created opportunities for private employers to divide trade unionists and environmentalists through "job blackmail" (Kazis and Grossman 1982; Loomis 2021: 129; Stevis 2019; Vachon 2023: 35). That is not to say that environmentalism is the sole object of job blackmail strategies. We hardly need to take a cursory look at today's headlines to see how some politicians use gender and other group characteristics to disrupt emancipatory politics. Instead, in recounting this particular moment in the history of labor environmentalism, we wish to underscore how transitional periods unfold through a series of political choices, and that political elites may use them to drive wedges between would-be allies and their issues. In any case, the hegemonic "jobs versus environment" frame has haunted labor environmentalism ever since.

During the late 1970s and early 1980s, groups such as Environmentalists for Full Employment sought to address this narrative by advocating that the country could achieve full employment by moving away from fossil fuels and nuclear energy—a major disagreement among unions—to alternative energy sources such as solar and wind, voicing many of the same arguments for these technologies still used today (Gordon 2004, Chapter 7; Grossman and Daneker 1979; Hultgren 2025)[1]. Despite some of these early instances of labor environmentalism and alliances between labor and the environment, labor environmentalism was more tempered in the 1980s as neoliberal policies took hold and the Reagan administration put workers and organized labor on its heels. Even still, events like the Black Lake conference hosted by the UAW and corresponding social democratic policies like the "freedom budget" showed the promise of a joint labor–environmental–civil rights coalition (Gordon 2004; Rector 2018). Moreover, workshop participants at the conference also discussed the potential of job creation in what many today would consider quintessential examples of "green jobs" in professions such as conservation, pollution control, and wind and solar power (Rector 2018). However, despite the promise of such efforts, unemployment, capital flight, austerity, and other structural forces made establishing a long-lasting coalition across social movements unlikely.

The complexity of labor and environment relations was evident in the spotted owl controversy that took center stage during the 1980s as union leadership of the IWA—once supporters of environmental protections—allied themselves with the timber industry, coming into conflict with environmentalists over the protection of old-growth forests where the spotted owl resides (Cooper 1992; Loomis 2015, 2021; Radkau 2014: 275–279). In popular consciousness, the spotted owl represents the quintessential example of the jobs versus environment frame in action. It also serves as the backdrop to a 1991 Worldwatch Institute report by Michael Renner titled "Jobs in a Sustainable Economy." The report begins with an oft-used bumper sticker that reads, "Save a logger. Kill an owl" (Renner 1991: 5), which Renner suggests is indicative of how many people, including politicians, often consider the relationship between economic and environmental issues (Renner 1991: 8). Like Grossman and Daneker (1979) and the emerging ecological modernization and sustainable development approaches, Renner suggested that the environment and *sustainability* can act as potential drivers for economic growth and innovation.

Efforts to combine environmental policy and economic growth/performance have their roots as early as the 1972 United Nations Conference on the Human Environment in Stockholm. Around that time, there was intense debate between the Global North and Global South over environment and development. An early response was that of ecodevelopment, an approach that was refined by the economist Ignacy Sachs (Monneret 2020). In the United States, Amory Lovins extolled the promise of green innovation and growth. Continental Europeans, some influenced by Lovins and Sachs, developed the concept of ecological modernization while the 1987 Brundtland Report[2] popularized the idea of sustainable development (i.e., paid work in formal arrangements that meet the needs of the present generation without compromising the capacities of future or foreign people to meet their needs). Renner's framing, as well as those by others during the 1990s, need to be understood in a longer context of attempts to fuse environment and development or growth. However, approaches varied a great deal, even within U.S. labor.

During the late 1980s, the OCAW developed the "superfund for workers" approach that soon morphed into the just transition strategy (Leopold 2007; Stevis 2019; Vachon 2023: 117). Jobs and green industrial policy were central to a just transition, but the opposition to just transition was quite strong in the United States. Some were in favor of a green transition (e.g., USW and even elements of the AFL-CIO; see United Steelworkers 1990 and Stevis 2021).[3] Other unions, particularly the UMW and the IBEW, were resolutely against any transition, particularly in the production of energy, which was becoming more prominent with the rise of climate politics. One reason offered is that it was too reactive and cast workers as supplicants—hardly in keeping with an ideological ethos of rugged individualism.

Another reason, however, was the fact that the just transition strategy was part of a broader democratic socialist strategy reflected in the program of the Labor Party that the OCAW sought to create—one which would have been well at home in Senator Bernie Sanders' political platform today—including things like single-payer healthcare and free college (Stevis 2021b: 597). Stated differently, just transition placed a broader political economy on the agenda of labor unions. A third reason had to do with the fact that the OCAW was a small union with internal problems that led to its absorption by an anti-environmentalist union, first, and the USW later. These tensions played out during the 1990s.

## Green Jobs With and Without Just Transition

The 1990s, in general, was a period of engagement between unions and environmentalists. One motivating issue was that of trade, exemplified by NAFTA and the 1999 WTO protests in Seattle (Dewey 2019). The USW and Sierra Club joined forces to argue for enforceable labor and environmental standards as part of these protests (Labor Network for Sustainability 2009) and established a relationship that continues today through the BlueGreen Alliance (BGA).

A key attempt to bring environmentalists together was the AFL-CIO's BlueGreen Working Group, which came about during a period of increased climate awareness (reflected in the 1997 Kyoto Protocol) and the aforementioned labor and environmental opposition to neoliberal trade policies like NAFTA and the WTO (Stevis 2021: 599). A central effort of the working group was to formulate a comprehensive labor–environmentalist agenda in which climate policy and just transition became central tenets. Early hopeful signs (e.g., Fellner 1998) were defeated by strong opposition to transition by energy unions and the slow uptake of just transition by most mainstream environmentalists. The final effort to salvage this initiative coincided with September 11, 2001. Needless to say, the meeting did not occur. Some working group efforts found their way into print, but the language of just transition was largely absent (Barret et al. 2002; see Renner 2000 for a full background). Nonetheless, discussions between some unions and some environmentalists—particularly those that resulted in the Apollo Alliance and the BGA—continued.[4]

September 11 was important for many reasons, including strengthening the already existing nationalistic tendencies within U.S. labor and expanding it into environmental employment. For instance, the Apollo Alliance (which would merge with the BGA in 2011) suggested that a shift in energy resources away from fossil fuels could simultaneously produce jobs, tackle climate issues, and enhance the country's energy security. A banner on their website from 2003 reads, "Three Million New Jobs. Freedom from Foreign Oil." Supported by key elements within the Democratic Party but also by environmental organizations and labor unions, Apollo offered a view of "sustainable employment" that combined environmental goals with economic and technological innovation (Little 2005; Stevis 2013). In keeping with weak technological modernization, they cited the potential for economic investments and technological innovation to spur job growth.

Another outcome of this period is the establishment of the BGA in 2006 (Mattera 2009: 10; Stevis 2021: 607). BGA was jointly established by labor and environmental groups—led by the USW and the Sierra Club—which gave it legitimacy from both camps. BGA also blazed its own trail through the adoption of an "all-of-the-above" approach to energy, but especially natural gas. Its position was that fracking could be achieved with minimal damage to the environment, so long as the necessary precautions are taken—and this natural gas would lessen U.S. dependence on foreign energy resources (Stevis 2019). The BGA, which absorbed the Apollo Alliance in 2011, increasingly coalesced around the idea of green industrial policy but one which did not include just transition. Others—like the Labor Network for Sustainability formed in 2007 with the leadership of Joe Uehlein, the ex-director of the AFL-CIO's Industrial Department—would chart an alternate direction, floating the notion of a just transition once again (Stevis 2021: 613).

## Green Jobs Before and Into the Financial Crisis

The strategy of green jobs as a form of sustainable employment and economic expansion had already become prominent outside of the United States, including Australia and the United Kingdom. In the mid-2000s, the Commisiones Obreras and their labor environmentalist NGO, Sustainalabour, as well as Cornell's Labor Institute, spearheaded a collaboration between the International Trade Union Confederation, the International Labour Organization (ILO), and the United Nations Environment Programme (UNEP) centered around green jobs. The lead author was Michael Renner, mentioned earlier. This resulted in a report titled, "Green Jobs: Towards Decent Work in a Sustainable, Low-Carbon World," which is described as "the first comprehensive report on the emergence of a 'green economy' and its impact on the world of work in the 21st century" (Renner, Sweeney, and Kubit 2008; United Nations Environment Programme 2008). Perhaps because the report is written from a global perspective, it considers the importance of tracing social and environmental impacts across borders and incorporating social and environmental full-cost pricing to discourage unsustainable patterns of production and consumption (Renner, Sweeney, and Kubit 2008: 4). The authors argue for the ILO's decent work agenda,[5] notably that social and environmental upgrading as a green economy strategy is *incompatible with economic policies* in which companies compete according to prices, as it invariably leads to the externalization of social and environmental impacts elsewhere as companies seek the cheapest inputs of materials and labor to remain competitive.

As the report was being finalized, the financial crisis began. As it wore on, labor–environment organizations increasingly made green jobs and green industrial policy central to the recovery (AFL-CIO 2009; Apollo Alliance 2008; BlueGreen Alliance 2011a; Good Jobs First 2009; see also Mattera 2009).[6] While Apollo was probably the most nationalistic, elements were present in all cases. BGA, for instance, responded to the 2008 financial crisis with a nationwide jobs plan to drive job creation through policies and investments into a sustainable and *growing* economy, stressing economic dimensions to keep workers employed (BlueGreenAlliance 2011a). The AFL-CIO even re-engaged with just transition, albeit on the green jobs side. It also allowed the International Trade Union Confederation to place just transition on the agenda of global climate negotiations.[7] Good Jobs First stressed that newly created jobs should be "good" to put the country on a high road to recovery (Mattera 2009). To do so, they suggested that strong labor rights and standards and increased union participation in middle-class sectors such as construction and manufacturing were necessary to improve work conditions for Americans across the board. It is worth noting, however, that they all stop short of calling for international agreements to establish global standards, despite a keen awareness of the issue of capital flight (see Mattera 2009: 19).

The Obama administration brought further attention to green jobs (Pollin et al. 2008). The American Reconstruction and Redevelopment Act of 2009 was designed to help reboot the economy in the throes of recession and included earmarks for renewables, electric vehicles, smart grid, and transit alongside other investment tax credits (Osaka 2020). On the heels of this, the U.S. Bureau of Labor Statistics (BLS) established the Green Jobs Initiative in 2010 to track green jobs throughout the economy. As part of this, the bureau defined green jobs as employment in "businesses that produce goods or provide services that benefit the environment or conserve natural resources, [or] jobs in which workers' duties involve making their establishment's production processes more environmentally friendly or use fewer natural resources" (U.S. Bureau of Labor Statistics 2011).

Established to assess policy initiatives and labor market impacts of economic activity for the purpose of environmental protection and conservation, the BLS noted that such information will be useful for the state, business, and job seekers.[8] By framing green jobs as those that "make processes more environmentally friendly" while upholding existing practices, the BLS view of green jobs is consistent with ecological modernization, as it advances a view of environmental work that conforms with existing economic practices—hardly a socially or transformative restructuring of the U.S. economy.

## Energy Back in the Mix

It is well known that U.S. labor unions lack the bargaining position of their peers in different parts of Europe. As a consequence, many support business unionism, meaning they focus primarily on projects that can deliver benefits to their members.[9] Even during the Obama administration, BGA and other groups tempered their expectations and messaging on "green jobs" at the national level as Republicans controlled Congress and the labor movement remained divided over the Keystone XL and Dakota Access pipelines (Stevis 2019: 6; Sweeney 2012). BGA and associated members such as the pipefitters and steelworker manufacturing unions advocated for an "all of the above" approach to employment in energy and environmental sectors in the mid-2010s, arguing that methane emissions and natural gas leakages could be prevented through responsible practices (Stevis 2019: 11). Similarly, the AFL-CIO supported (and continues to support) nuclear energy, an industry that increasingly markets itself as a clean, emissions-free energy source to provide bulk and baseload power needs for utilities (Savage and Soron 2011). Of course, while nuclear energy may limit emissions in the short run, it does little to consider the social and environmental consequences of nuclear waste in the long run or the potential impacts of mining to local communities (Höffken and Ramana 2024; Kyne and Bolin 2016).

The debates over the pipelines made clear that there are significant tensions over the green transition as well as the nature of green jobs. In the mid-2010s, natural gas had become a dominant energy source in the country, brought about by technological innovation in fracking that led to a 5.1%-per-year average growth rate between 2015 and 2020 (U.S. Energy Information Administration 2020: 46). In many respects, it is natural gas, not renewables, that in combination with increasing automation, has contributed to the rapid shuttering of coal power plants across the country, as they simply cannot compete with natural gas production based on cost. However, unlike wind and solar, natural gas does not lead to net reductions in climate emissions. The consequence of this are divisions within the labor movement, with some labor unions in support of the potential job opportunities within the industry, whereas others believe natural gas does not go far enough in achieving progress on state- and national-level renewable energy portfolio standards.

Increasingly, those more concerned about climate change sought another narrative. Writing from outside the United States, Jonathan Neale (2008), to our knowledge, offers the first concrete differentiation between green jobs and climate jobs. In the report, "One Million Jobs," Neale argues that climate jobs are those that focus on emissions reductions while making communities more resilient to climate change (Neale 2014: 4). Written at a time of ascendent discussions on climate justice, climate jobs represent a smaller slice of "green jobs," with a narrower focus on addressing climate issues and arguably greater attention to social justice issues.

Domestically, the idea of climate jobs appears to have gained traction with Climate Jobs New York, which began in 2017 as a "labor-led coalition that works to combat climate change, create good union jobs, and reverse racial and economic inequality by building a worker-centered clean energy economy" (Climate Jobs National Resource Center 2024). Superstorm Sandy inspired the establishment of Climate Jobs New York, as the hurricane cost upward of $88.5 billion according to NOAA estimates (National Centers for Environmental Information 2024a, 2025b). Climate Jobs New York began as a coalition of groups including District 37 of the American Federation of State County and Municipal Employees and the Building & Construction Trades Council of Greater New York (BCTC), that were more invested in climate change mitigation efforts, and had early successes in securing climate investments in offshore wind and working to pass state-based legislation to advance investments into lower-carbon infrastructure. Later in 2020, Climate Jobs New York joined with the Climate Jobs Institute to form the Climate Jobs National Resource Center (CJNRC) and CJNRC Action Fund to advance a climate work agenda across the country. CJNRC differentiates itself from BGA with membership solely comprised of labor unions and councils, largely from the building and construction trades, groups that have been historically skeptical about the green transition and, even more so, a just transition. The organization is making important progress in redirecting these unions, although just transition appears even less prominent than it does with the BGA. CJNRC and its local affiliates have helped pass legislation such as a 2021 New York budget bill that included prevailing wage and project labor agreement requirements for construction on renewable-energy projects of five megawatts or greater, in addition to sourcing American steel and iron where feasible (ILR Climate Jobs Institute 2021). Interestingly, the CJNRC expanded what climate jobs might look like. In a series of vignettes, they explore different climate jobs, and these include stories from educators and IT technicians. On the surface, these professions have seemingly little to do with climate issues. However, dig a little bit deeper and we can understand their climate nexus. For example, the IT technician provides a critical service to the community by ensuring residents, including schoolchildren, can access the Internet and all the opportunities that entails. In addition to reducing pollution associated with transportation, the technician's work simultaneously ensures more equitable access to information, making this form of work more just and human centered. Another vignette explores the role teachers play in educating the next generation about environmental issues helping the school district and community pursue locally owned renewable energy systems for the school (Vachon 2023). This move by CJNRC suggests that we broaden our lens to consider how all forms of work—waged and nonwaged—might help us move in more socially and ecologically sensitive directions.

## A Green New Deal

The tensions within the world of labor were clearly reflected in their response to the Green New Deal (GND). In 2018, Representative Alexandria Ocasio-Cortez and Senator Ed Markey introduced a GND resolution to Congress to advance principles of jobs, justice, and climate action (Ocasio-Cortez 2019).[10] The GND put a climate spin on an industrial policy agenda that harkens back to President Franklin Roosevelt's federal government works and conservation programs during the dregs of the Great Depression.[11] The GND resolution sent ripples throughout the country by calling for nationwide mobilization to achieve decarbonization through 100% renewable energy power generation and modernization of grid infrastructure. Like FDR's New Deal programs, the GND recognizes that larger-scale efforts are necessary to transform the economy so that it provides material security to all, but especially through the creation of millions of high-wage jobs. Unsurprisingly, the resolution's emphasis on economic change and reductions to inequality led many conservative pundits and politicians to label it as socialist propaganda masquerading as climate activism.

The AFL-CIO formally came out against the measure, citing potential job losses to some of its union members, as well as resentment for not having been included in drafting the resolution (Roberts and Stephenson 2019). Other unions who signed in opposition included the UMWA, IBEW, LIUNA, UA, USW, IBB, IW, UWUA, IUOE, and NABTU. Both the United Mine Workers of America and International Brotherhood of Boilermakers criticized the GND resolution for not approaching the transition with "rational policy." Both organizations recognize the presence of an unfolding transition but remain skeptical it can occur in a "just" manner. In both cases, the organizations advocate for carbon capture use and storage technologies (CCUS), measures that attempt to remedy ecological harms in a post-hoc fashion (Jones 2019; Young 2021).

Despite the opposition to the GND resolution, it helped expand what elements of labor and environmentalists believed to be possible (Brecher 2024). Even still, many local unions and affiliates with the AFL-CIO, like the AFT and CWA, supported the GND resolution. Moreover, several organizations published their own vision of a GND. BGA members who initially opposed the GND resolution—such as the UWUA, UA, and USW—would later support a BGA report titled *Solidarity for Climate Action*, written in reference to the initial GND resolution. The BGA report also argues that "ironclad" protections are necessary to protect workers across the supply chains of an industry, an important extension of ethical considerations across space. BGA's report also calls for guaranteed pensions and social security nets for workers impacted by the transition. Even still, the BGA report champions technological fixes and associated employment as the primary way to achieve emissions reductions. and it continues to advocate for an "all-of-the-above" approach that includes things like carbon capture, removal, storage, and nuclear (BlueGreen Alliance 2019: 4).

Others, such as the Labor Network for Sustainability, published the aptly titled "18 Strategies for a Green New Deal" (Brecher 2019), which presents one of the clearest depictions of a transformative vision of a new energy economy painted with darker green hues. The report has several key provisions, like a jobs guarantee, the protection of worker rights, and the elevation of unions as vanguards for infrastructure build-out. It recognizes that jobs are embedded in a broader political economy and calls for binding international agreements to ensure the transition occurs equitably across borders, which includes knowledge-sharing to make the transition achievable in both the Global North and South. At the heart of the report lies recognition that the hegemonic neoliberal policies that helped accelerate globalization and environmental damage do not serve Americans, and fundamental changes cannot be achieved through technological innovation alone. As part of this, there is greater emphasis on questions of power and control in a "new energy economy," including the manner in which energy is produced, consumed and distributed—each of which are fundamentally political—even though technologies might change the scope of what is possible.

## COVID Aftermath, the IRA, and Mainstreaming of Green Jobs and Climate Jobs

In the midst of the COVID-19 pandemic, earlier efforts of labor–climate organizations (including the BGA and the CJNRC, who worked to promote the idea of green industrial policy) bore fruit. In August 2022, Congress passed the Inflation Reduction Act (IRA). The law includes many policies designed to help accelerate investment into low-carbon technologies and manufacturing, supplementing other acts such as the Infrastructure Investment and Jobs Act (2021) and CHIPS and Science Act (2022). Early reports indicate that the IRA alone may add nearly 1.5 million jobs to the U.S. economy by 2030 (Foster, Maranville, and Savitz 2023).

IRA impact estimates vary widely because of the broad-based set of incentives and heavy use of tax credits throughout the law, credits that include multipliers for projects in qualifying "energy" or "low-income" communities. Regardless, actual climate spending could be much higher than the Congressional Budget Office (CBO) estimates pegged the number at about $391 billion, with others such as Goldman Sachs putting the number closer to $1.2 trillion (Saul 2023). IRA incentives included provisions that stipulate that new energy projects must fulfill Fair Labor Standards Act (FLSA) requirements, uphold the Davis–Bacon Act to pay prevailing wage provisions and use a certain percentage of registered apprentices to receive tax credits. The associated tax credits cover 30 percent of total project costs if the employer pays the prevailing wage on a project and meets an apprenticeship ratio of 10% in 2023 and 15% by 2024 (Manzo, Wilson, and Kashian 2022: 14). Finally, the IRA included incentives to improve racial equity and better help underserved communities as part of President Biden's Justice40 initiative that sought to deliver 40% of these investments into disadvantaged communities (U.S. Department of Energy 2021).

While the tax credits and other incentives in the IRA are a positive signal of intent, their impact on quality of work remains to be seen and now, following the 2024 election, we may never know. Prevailing wages can be useful, but they do not ensure workers receive crucial benefits such as retirement, healthcare coverage, or family and sick leave. What is more, prevailing wages may still be inadequate to provide sufficient economic security to workers. The IRA specifies that failure to comply with prevailing wage and apprenticeship standards in claiming tax credits makes the company liable for back wages and IRS penalties. However, these amount to little more than a mild reprimand when it comes to enforcement. Many companies may choose to leverage subcontractors in their production networks to keep costs down on projects (Gadzanku, Kramer, and Smith 2023; Mayers 2024).[12]

Few measures in these bills seek direct emissions reductions, instead favoring tax credits and other incentives to supercharge investments into the expanding new energy economy. Rather than reconstitute environmental and economic relations, these continue to reinforce the underlying logic of capital expansion. When the legislation directly invokes workers, it often emphasizes the importance of social reproduction—training individuals to create a pipeline of workers. Many reports from the U.S. government, including the Department of Energy and affiliated laboratories, describe employment in the energy transition this way.

Also of note, President Biden launched the American Climate Corps through executive action in September 2023 to get young people trained into climate and environmentally focused careers (White House 2023). As part of Biden's Justice40 efforts, the initiative includes projects beyond energy that look to conserve and restore lands and waters, including in environmental justice communities. Like the IRA, this effort is significant but hardly transformational. In a cursory scan of the potential employment opportunities on the Climate Corps website during the Biden administration, many positions were nonunion, and many of the positions pay wages that fall far short of the living wage requirements where these jobs exist (how feasible is a part-time job that pays $20/hour in San Francisco?). Moreover, there is little oversight to enforce or oversee that the Justice40 initiative actually directs a flow of funding and hiring to disadvantaged communities within the American Climate Corps. By contrast, the Civilian Conservation Corps, to which the Climate Corps owes inspiration, put millions of Americans to work and was instrumental in helping shift American attitudes toward the environment, tilting it to a more conservation-minded ethic (Maher 2009).[13] In any case, the American Climate Corps was not a relief program; instead, it focused on providing green businesses and industries the workers they needed to make progress on Biden's industrial agenda.

The IRA, Climate Corps, and other policies had more limited provisions to support communities in transition. Some funding has been carved out for communities in transition, and the Interagency Working Group on Coal and Plant Communities was set up to assist communities with economic revitalization efforts. Despite this, these policies had few explicit just-transition components that would have expanded the social safety net for workers and communities. Even within the labor–climate movement, however, advocacy for a just transition has been, unsurprisingly, somewhat tepid (Stevis 2021). With Trump back in office, many initiatives like the Interagency Working Group have stopped meeting—after all, acknowledging the replacement of coal by natural gas and renewables does not fit the administration's narrative. The president has promised to "unleash American energy" but has made permitting for new solar and wind projects far more onerous, and effectively gutted the subsidies put in place by the IRA—impacts that will not only reduce the number of jobs available to workers, but also the prospective quality of these jobs as companies look for ways to maintain their viability in the face of the added challenge.

## A TYPOLOGY OF GREEN JOBS AND CLIMATE JOBS

As the preceding history demonstrates, notions of sustainable work have evolved considerably, often according to important historical turns. For example, green jobs may or may not be connected with green industrial policy or just transitions, and they may not be the same as climate jobs. In each case, how these expressions are defined reflects different visions of work in a sustainable economy—imaginings that serve different interests and ends according to the political actors that deploy these terms. Like all important concepts, green jobs and climate jobs are "essentially contested," meaning that they involve some intersubjective language but not agreement on the definition (see Connely 2007 and Gallie 1955–1956). Moreover, technical differences do not create disagreement, but rather operationalizations embedded within particular ideological/political economy preferences. The first step, therefore, is not to push for a superior or common definition. Rather, each entity that uses the terms should state them as clearly as possible (see Connelly 2007).

To help us understand the history we have constructed and consider alternative visions of what "sustainable work might entail," we turn to an earlier typology created by Kate Crowley (1999).[14] Writing from Australia, where ecological modernization (EM)[15] grew in popularity during the 1990s, Crowley sought to understand whether employment opportunities associated with Australian EM policies had utility in overcoming the "labor versus environment" binary, and if they would alter social relations between Australian political economy and the environment. Building on earlier work (Crowley 1996a, 1996b), Crowley concludes that a "truly green jobs agenda" would address ecological decline through ecological rationality, authenticity, and modernity, but that such visions remain elusive, even though EM offers a "pragmatic institutional pathway for *greener* employment outcomes" by limiting or eliminating environmental impacts associated with economic processes (Crowley 1999: 1014; emphasis added). Others take this idea further in the belief that decoupling environmental impacts from economic processes is not only possible but may enhance economic growth, a viewpoint known as "green growth" (Crowley 1999; Kouri and Clark 2014). Such policies often focus on increasing employment for industries (i.e., the social reproductive aspects of industry).

Crowley's typology articulates the contours of different approaches to green employment with a series of descriptors that assess socioenvironmental relations implicit in these views. She classifies different visions of green jobs in a green economy as light, mid, or deep green. The typology helps make explicit how certain attributes of green jobs/climate jobs contribute to more or less environmentally transformative visions of employment according to the extent to which they change societal relations with the environment, with deep green visions requiring *fundamental* ecological improvements that redefine growth and enhance ecological sustainability (Crowley 1999: 1108). Contrast this with light green jobs that largely react to environmental issues and regulation, or mid green visions that look to balance economic and ecological priorities by making existing economic practices greener.[16]

While there is much to like about Crowley's typology, it can be difficult analytically to disentangle social and ecological dimensions within it. After all, a "green job" can have deep or mid ecological aspects with respect to environmental outcomes and processes but still offer insufficient social welfare to the individual who occupies the position. Similarly, employment policies may come with social protections and benefits but actively contribute to the destruction of the planet and its people. To make these social dimensions explicit, we add several factors to better capture social dimensions of work (Table 1). These include nonprovincial/relational justice, social objective, International Labour Organization (ILO) labor rights and standards, and social aim.

*Nonprovincial, relational thinking* recognizes the mutual constitution of eco-social relations across space. It covers both blue/social and green/environmental aspects with awareness for how practices in one space may rely on socially and environmentally harmful practices elsewhere. It is important to recognize the processual aspects of work that may produce green outcomes but do so through socially and environmentally harmful processes (Stevis 2013, 2023; Van Der Re 2019).

The ILO category considers whether green jobs/climate jobs policy platforms are "decent" and "good" in contributing to socially transformative progress by ensuring the provision of a fair income, security, and social protection in the workplace; better prospects for personal development and social integration; freedom to express concerns; the ability to collectively organize and participate in the decisions that affect their lives; and equality of opportunity and treatment for all individuals.

Table 1. Modified Green Jobs Typology

| | Deep green | Mid green | Light green |
|---|---|---|---|
| Mode | Proactive | Integrative | Reactive |
| Scope | Long term | Intermediate term | Short term |
| Nature | Transforming | Reforming | Conforming |
| Objective | Redefine growth | "Ecologize" growth | Enhance growth |
| Operation | Rejectionist | Reinventionist | Accommodationist |
| Aim | Ecological sustainability | Ecological modernity | Sustainable development |
| Jobs | Preserving nature | Greening industry | Remedying ecological decline |

| | Deep blue | Mid blue | Light blue |
|---|---|---|---|
| Social objective | Human centered | "Humanize" growth | "Humanize" growth |
| ILO labor rights and standards | Upholds labor rights and standards across global production networks | Uphold labor rights and standards at site of production | Does not respect ILO labor rights and standards |
| Social aim | Social transformation | Social reform | Status quo |

| *Nonprovincial, relational justice | Ethical considerations extend across space and time | Ethical considerations extend to group or nation | Limits to ethical considerations |
|---|---|---|---|

*The bottom row makes more explicit social dimensions of green jobs/climate jobs. The original table items in Crowley (1999) are the rows in bold type.

The social objective category judges the extent to which workers pursue (or are empowered to pursue) human-centered economic activity. Deep blue positions are those that build up communities and individuals who move away from profit-oriented, destructive, and extractive enterprises to human-centered activity. Moreover, deep blue positions affirm for the individual who performs work that their activities are socially useful, in contrast to positions that solely exist for profit making or work for its own sake.

Finally, the social aim category synthesizes the others and considers whether a particular view of green jobs/climate jobs is socially desirable and ethically defensible. Deep blue jobs or visions aim for societal transformation that redefines social relations in political economy, which we understand as fundamentally altering what aspects of work are valued, collective empowerment, and how these might better serve communities and, by extension, the environment.

Salient incarnations of what we consider a deep green/blue job—jobs that are deep along both ecological and social dimensions—might include visions that would have been at home in the 1976 Lucas Plan[17] or advocated for by today's Just Transition Alliance.[18] In both cases, the organizations reappraise the role work plays in the economy, including what ends economic activity serves. The reappraisals aim at fundamental transformation through a revaluation of the role work and workers play in society, including how workers relate to one another and more-than-human ecologies across space and time. For our purposes, we refer to such visions as "deep eco-social" visions of work for their ability to integrate ecological, social, and spatial aspects of work in the economy simultaneously.

## REFLECTIONS ON GREEN JOBS AND CLIMATE JOBS

In light of our typology and the history we have constructed, what can we learn? Starting with the most basic observation, groups use salient political events to frame their viewpoints on green jobs/climate jobs. Organizations infrequently discuss jobs or work without acknowledgement of the broader political economy in which it is embedded. To discuss a green job necessarily brings with it a normative vision of what the economy ought to look like, as green jobs exist in juxtaposition to the other forms of work that are considered standard or the norm in society.

Few organizations or trade unions have held deep green ecological views. Perhaps unsurprisingly, most groups have framed sustainable employment for its potential to benefit the economy while also addressing environmental issues (and climate emissions, in particular). Organizations often reasoned a "double dividend" could benefit both the economy and environment. Many of these emphasized that markets, technology, and efficiency gains would all produce employment. The promotion of green jobs/climate jobs in these views is explicitly linked to the development of new green industries that are being driven by capital investments by U.S. state and financial sectors under the auspices of environmental sustainability; in other words, a prototypically ecologically modernist approach to employment and the environment— one that emphasizes continual capital growth and the expansion of the economy. Such efforts may help overcome the apparent jobs-versus-environment divide, but they do not necessarily reconstitute or shift social and environmental relations in a direction that simultaneously protects the environment and empowers workers and communities.

However, we have seen some labor unions and groups aligned with the labor–climate movement (LCM) advocate for more socially transformative work arrangements. For instance, Good Jobs First (Mattera 2009) emphasizes that employment should make the lives of working people more secure. Earlier, Mazzocchi advocated for a "worker superfund" to help individuals and communities cope with transitions to the economy— indeed, he reasoned the government should pay workers *not to work* to avoid environmental destruction. While this may seem like a marginal difference from mid green/blue visions, it has consequential policy implications for labor during an energy transition. One vision caters to businesses by asking it to "green" their production processes, which has the knock-on consequence of creating new employment opportunities. Many of these are engendered by public policies like the IRA, which was designed to create a pipeline of workers *for* industries. By contrast, social imaginaries put forth by groups like Mazzocchi or the Just Transition Alliance center justice and job quality, and better incorporate broader economic change into their political platforms of work. In so doing, they propose a fundamental shift in societal value relations. They consider how waged labor, and in some instances non-waged labor, might act as the vanguard for more just communities. After all, work is a near universal human experience and the arena where most of us spend the majority of our waking hours.

To acknowledge this is to also recognize that employment has become increasingly precarious as unionization has declined and "gigification" of the economy proliferates (Luce 2022). Federal and state governments have an important role to play in the social upgrading of work, as businesses will not create "good green jobs" of their own volition. However, in the absence of support from the federal government and many state governments, it has (once again) fallen to workers themselves to make new jobs in green industries good jobs. Groups such as Green Workers Alliance have worked diligently to make inroads among solar and wind energy workers to improve work conditions. As they stand now, green industries are no different than other sectors with respect to their goals and purposes. However, these industries and changes to the energy sector more broadly differ in the potential they bring to create political change—energy is nothing if not political, and history demonstrates how workers in energy sectors or other "strategic locations" at the heart of the world economy are positioned to create change that reverberate far outside the confines of their individual workplaces (Braverman 1998; Malm 2016; Mitchell 2011; Silver 2003).

## (DEEP) ECO-SOCIALLY INTEGRATIVE VISIONS OF WORK

Deep eco-social visions of work ask us to expand the aperture of how we think about work and employment in our lives and communities. For instance, an organization such as Familias Unidas por la Justicia recognizes the central role migrant workers play in contemporary industrial-scale agriculture, an industry implicated in the climate crisis that relies on ecologically and socially exploitative practices for profit. However, Familias has made inroads in organizing migrant farmworkers—despite their lack of coverage by the NLRA—successfully earning their first contract through organizing a public boycott of Driscoll's (Ordower 2024: 30). Beyond this important organizing effort, what makes the group transformational is their pivot toward communal ownership of land to enact regenerative growing practices with native plants and green agricultural processes and make agribusiness less exploitative while simultaneously uplifting work conditions in the industry (Ordower 2024).

Deep eco-social visions move the conversation about work and workers away from profit-making and productive enterprise to a movement centered on revaluation of both work and the environment. Deep eco-social visions recognize that the fate of workers and the environment are inextricably connected, and this entails recognition of cross-spatial and temporal considerations that an eco-socially integrative economy is *necessarily transformational in both social and environmental realms.*

We are hardly the first to make such a case. African American activist William B. Ratliff of the South Carolina Urban League made a similar, far more eloquent argument, remarking, "The basic causes of environmental and economic injustice [are] discrimination, over-concentration of wealth, and a shortage of economic democracy" (Rector 2018). Ratliff would also argue that the sphere of social reproduction and the provision of social democratic policies like full employment could better enable unions and civil rights organizations to collaborate with environmentalists (Rector 2018).

Underlying these deep eco-social visions is recognition that the existing political–economic system—based on growth and consumption for its own sake—does not serve the future well-being of people and the planet. Jobs, or even work more generally, must undergo qualitative changes if we are to make a transition with justice that shows new employment forms are not only possible but better than what currently exists. However, this requires us to remain attentive to more than just net job gains. It requires us to consider the broader political economy in which jobs—and work more generally—are embedded within, which makes these green jobs or climate jobs more or less socially sustainable. What we have done here is to track how these frames and associated visions of work have changed throughout the country's history.

Of course, this raises the question as to whether deep green visions are possible within the current architecture of U.S. society and global capitalism. LCM-aligned organizations and trade unions are well aware of the dynamics in which they operate, and this influences how they think about employment in a green economy (Stevis 2019; Vachon 2023). As we saw in the case of union divisions over the Keystone XL pipeline, unions may support a business position if it helps secure employment for their members.

In the United States, the tables are tilted in favor of business when it comes to social dialogue around labor and environmental issues, which reinforces what groups believe to be politically feasible. Even under Biden, arguably the most pro-union administration since FDR, the litany of tax credits and investments available to corporations in the IRA, IIJA, and CHIPS and Science Act lets the hand slip—businesses and private utilities are the primary beneficiaries of these policies. Combined with a regulatory environment hostile to organized labor and a social welfare regime that attaches healthcare to employment, the willingness and possibility for unions and other LCM groups to demand more transformative changes is significantly curtailed. As a result, many organizations encountered in constructing this history pursue strategies that amount to tweaks around the edges to withdraw concessions from the political process rather than a reconstitution of social and environmental relations.

While this is understandable given the society in which they are embedded, if we read labor history as an ever-expanding tent of concern, then future visions for deep eco-social jobs will require us to rethink what social solidarity looks like. More specifically, it may require a recollection of the most nonprovincial and encompassing ideas during the heroic age of labor[19] that lie dormant in the collective memory of U.S. labor, as well as an extension of the notion of justice to other communities, including future generations and the planet (Pederson et al. 2024).

## FINAL THOUGHTS

To our mind, the recent election of Donald Trump does little to slow the ongoing energy transition globally, though it certainly changes its trajectory and character in the US. Indeed, green jobs and climate jobs as labels have become something of a liability in the early days of the Trump administration. Despite the bleak outlook for energy transition efforts in the US, the authoritarian turn at the federal level also engenders urgent calls for more radical positions and possibilities for what work and energy systems could look like.

In some ways, we are seeing, again, those more innovative efforts develop at the state and local level across parts of the country (Brecher 2024). However, there is a real danger that if newly created positions within the new energy economy are not good, decent jobs, they may only serve to further exacerbate precarity and the politics of resentment that create fertile conditions for politicians like President Trump to take root. There is a clear need to reconceive eco-social relations, and if the waves of protests over the initial months of President Trump's administration are any indication, there is also an outward desire exists to do things differently. Though deep eco-social visions of work in the economy have been advanced, more groups and trade unions have adopted a view of green jobs and climate jobs consistent with ecological modernization—that is, a view of sustainable work that continues to prioritize business and capital interests with green varnish.

As we have argued, these will not suffice.

## ENDNOTES

1. An oft-referenced reason for shifting to alternative energy technologies is their potential to increase employment. For instance, Richard Grossman and Gail Daneker (1979: 79) argue that solar development requires two and a half jobs more than the same amount of power produced from nuclear fission.

2. The World Commission on Environment and Development produced a report in 1987 titled "Our Common Future" (also known as the Brundtland Report for its principal architect) that popularized the concept of sustainable development (Carter 2018: 213). Sustainable development was crafted as a "bridging concept" to temper the tensions between environmental and development (similar to jobs versus environment) that had become an important South–North issue since the very early 1970s (Meadowcroft 2009).

3. In 1990, the United Steelworkers published "Our Children's Future," calling attention to several environmental issues and the imperative for the union to preserve and protect their children's future. The report cuts through the mirage of jobs versus the environment and makes a case for how production processes can be "greened," offering an intermediate or long-term view of what the future might look like—flagging that the issues are "not technical, but political" in nature.

4. The debate on jobs and environment was not limited to the negotiations between unions and environmentalists. For example, see Goodstein (1999). In general, a wide range of think tanks had been converging around some form of ecological modernization—without using the term— for many decades. These include Resources for the Future (1952–), World Resources Institute (1982–), Rocky Mountain Institute (1982–), and others.

5. Both the ITUC and ILO make the case that green jobs are "decent jobs," meaning that they provide "opportunities for work that is productive and delivers a fair income, security in the workplace and social protection for all, better prospects for personal development and social integration, freedom for people to express their concerns, organize and participate in the decisions that affect their lives and equality of opportunity and treatment for all women and men" (International Labour Organization 2013).

6. Parallel to these organizations, there was increasing academic research, both theoretical and applied (e.g., Block 2011; Pollin 2009: 13; Stevis 2013, 2014).

7. It is worth noting, however, that almost none of the Green New Deal proposals created during the crisis include just transition as a detailed and central component, even though crises and periods of economic change are essential components of the global economic order.

8. The BLS stopped tracking green jobs in 2013 due to budget cuts. Groups such as Data for Progress argue that a proper accounting of work in the green economy requires the BLS to reinstate its green jobs tracking initiative (Carlock, Mangan, and McElwee 2019: 17).

9. Many unions in the United States espouse an ethos consistent with business unionism, which sees a union's role as primarily in direct service to its members to help negotiate contracts and resolve business disputes with management. Contrast this with social unionism, which views unions as potential vessels for more widespread societal transformation, including improving working conditions (Burawoy 2008).

10. It should be noted that a Green New Deal task force entertained the idea in 2006 (Stewart 2018).

11. Mark Twain was actually the first to introduce the idea of a "new deal" as a call to reduce inequality and economic volatility in the gilded age (Calhoun and Fong 2022: 11). As the expression goes, history does not repeat itself, but it rhymes—a quote often attributed to Twain.

12. On this front, the IRA also includes new measures that encourage the use of project labor agreements (PLAs) and community workforce agreements (CWAs). Unique to the construction industry, PLAs establish terms and conditions of employment between one or more unions and one or more employers on a specific project (29 U.S.C. § 158 (f)). PLAs in the IRA are used to help employers secure tax benefits by ensuring compliance with the IRA's prevailing wage and apprenticeship requirements. The IRS would waive potential penalties for employers who have a qualifying PLA in place, so long as back wages and interest owed to laborers and mechanics are paid by the day they file for the increased credit (U.S. Department of Labor 2023). In any case, these provisions signaled the Biden administration's interest in providing a supply of low-carbon workers to achieve decarbonization efforts.

13. Labor unions did not support the CCC initially, believing that CCC jobs would undercut them. However, the AFL later came to support the CCC, recognizing it as an important relief program for the country. By contrast, environmentalists such as Aldo Leopold were initially supportive of the CCC (he even helped lead one of its regional offices) but increasingly took issue with the program's emphasis on land use for economic productivity rather than for its intrinsic properties (Maher 2009).

14. It should be noted that this modified table and the typology we have introduced here are early versions of a much more detailed and fully articulated typology we are continuing to develop.

15. Ecological modernist (EM) policy accepts that capitalism and associated economic processes produce social and environmental problems. In its "strong" form, it has many similarities with environmental sustainability in its attempt to combine social goals with environmental priorities, albeit with an added emphasis on business practices (Dryzek 2003: 167). In its "weak" form,

however, EM represents a "techno-corporatist" approach to environmental issues, as it remains transfixed on industrialized business processes and their response to the environment rather than on humanity's relationship with the environment (Carter 2018). This happens through a two-pronged effort of dematerialization (i.e., producing the same unit of economic output with fewer environmental inputs) and decoupling (i.e., loosening the connection between continued economic prosperity and environmental degradation) (Carter 2018). Ecological modernization offers the promise of a win–win, for both society and the environment (Barca 2020).

16. To provide some examples, light green employment policies might advance something like "clean coal," mid green policies would be those that seek to expand renewable energies and associated work, and deep green visions include things like regenerative local agriculture that seek to preserve nature and consider ecological impacts with extended time horizons.

17. The Lucas Plan was launched in 1976 by shop stewards produced by workers of the Lucas Aerospace Corporation in response to announced job cuts. In their plan, workers set about changing the workshop from one focused on crafting tools of terror and war to socially useful, human-centered products like wind turbines and heat pumps: a fundamental restructuring of the workplace, including democratizing technological development in society (Smith 2014). As Adrian Smith recounts, the Lucas Plan received a great deal of international attention (including a Nobel Peace prize nomination, no less), but the plan in isolation would not convince management to change or the government to intervene. Leaders in established trade unions were similarly reluctant to support the initiative, anxious about how it might unsettle existing demarcations and hierarchies.

18. The Just Transition Alliance (JTA) has come out with very tepid support for Biden administration policies like the IRA and Justice40 initiative. Where the Lucas Plan is deep in the social sense, JTA offers a deep green view of ecology, with efforts to make local communities cleaner and more energy sovereign (though we could not find a concrete plan of what that looks like) and criticism of proposed "clean energy solutions" such as the production of biodiesel or the expansion of carbon trading schemes that simply allow the unrestrained development of economic processes that sap energy resources.

19. The "heroic age of labor" in the United States falls roughly between 1865 and 1893, with groups such as the Knights of Labor, who sought to build solidarity across working people, including in their ranks workers across industries regardless of race, gender, or religion—although the group supported the Chinese Exclusion Act of 1882, which is a mark against their solidaristic pedigree (Nicholson 2014). Regardless, the group was instrumental in the promotion of the eight-hour workday until the Haymarket Riot of 1886 curtailed its influence.

## ACKNOWLEDGMENTS

The chapter authors would like to thank the following individuals who reviewed the chapter and provided feedback, which made for a far more robust and detailed history: John Hultgren (Bennington College), Erik Loomis (The University of Rhode Island), and Scott Dewey (University of Minnesota).

## REFERENCES

AFL-CIO. 2009. "Resolution 10: Creating and Retaining Sustainable Good Green Jobs." https://tinyurl.com/ammnnyd9

Apollo Alliance. 2008. "Our Mission." https://tinyurl.com/477hxc5m

Barrett, James, Andrew Hoerner, Steve Bernow, and Bill Dougherty. 2002. "Clean Energy and Jobs: A Comprehensive Approach to Climate Change and Energy Policy." Economic Policy Institute. https://tinyurl.com/yc27crdx

Barca, Stefania. 2019. "Labour and the Ecological Crisis: The Eco-Modernist Dilemma in Western Marxism(s) (1970s–2000s)." *Geoforum* 98 (January): 226–235. https://doi.org/10.1016/j.geoforum.2017.07.011

Block, Fred L., and Matthew R. Keller, eds. 2011. *State of Innovation: The U.S. Government's Role in Technology Development.* Paradigm Publishers.

BlueGreen Alliance. 2011a. "JOBS21! Good Jobs for the 21st Century." BlueGreen Alliance. https://tinyurl.com/nhfc7xtj

BlueGreen Alliance. 2011b. "Statement from BlueGreen Alliance's David Foster on BlueGreen, Apollo Alliance Merger." https://tinyurl.com/4u4v76rw

BlueGreen Alliance. 2019. "Solidarity for Climate Action." BlueGreen Alliance. https://tinyurl.com/yycve345

BlueGreen Alliance. 2024. https://www.bluegreenalliance.org

Bookchin, Murray. 1962. *Our Synthetic Environment.* Ig Publishing.

Bowen, Alex, Karlygash Kuralbayeva, and Eileen L. Tipoe. 2018. "Characterising Green Employment: The Impacts of 'Greening' on Workforce Composition." *Energy Economics* 72 (May): 263–75. https://doi.org/10.1016/j.eneco.2018.03.015

Braverman, Harry. 1998. *Labor and Monopoly Capital: The Degradation of Work in the Twentieth Century.* 25th anniversary edition. Monthly Review Press.

Brecher, Jeremy. 2019. "18 Strategies for a Green New Deal: How to Make Climate Mobilization Work." NGO Report. Labor Network for Sustainability. https://tinyurl.com/36mra7nc

Brecher, Jeremy. 2024. *The Green New Deal from Below: How Ordinary People Are Building a Just and Climate-Safe Economy.* University of Illinois Press.

Bresnihan, Patrick, and Naomi Millner. 2023. *All We Want Is the Earth: Land, Labour and Movements Beyond Environmentalism.* Bristol University Press.

Brown, Wendy. 2019. *In the Ruins of Neoliberalism: The Rise of Antidemocratic Politics in the West.* The Wellek Library Lectures. Columbia University Press.

Burawoy, Michael. 2008. "The Public Turn: From Labor Process to Labor Movement." *Work and Occupations* 35 (4): 371–387. https://doi.org/10.1177/0730888408325125

Calhoun, Craig, and Benjamin Fong. 2022. *The Green New Deal and the Future of Work.* Columbia University Press.

Carlock, Greg, Emily Mangan, and Sean McElwee. 2019. "A Green New Deal: A Progressive Vision for Environmental Sustainability and Economic Stability." Data for Progress. https://tinyurl.com/hbeewej7

Carter, Neil. 2018. *The Politics of the Environment: Ideas, Activism, Policy.* Third edition. Cambridge University Press.

Carson, Rachel. 2002. *Silent Spring.* 50th anniversary edition. Mariner Books, Houghton Mifflin Harcourt.

Clean Investment Monitor. 2024. "Overview of Clean Investment in the U.S." https://www.cleaninvestmentmonitor.org/database

Climate Jobs National Resource Center. 2024. "About Us." https://www.cjnrc.org/about-us

Cooper, Mary H. 1992. "Jobs vs. Environment." *CQ Researcher* 2 (18): 409–432.

Connelly, Steven. 2007. "Mapping Sustainable Development as a Contested Concept." *Local Environment* 12 (3): 259–278.

Cramer, Katherine J. 2016. *The Politics of Resentment: Rural Consciousness in Wisconsin and the Rise of Scott Walker.* Chicago Studies in American Politics. University of Chicago Press.

Crowley, Kate. 1996a. "Environmental Employment Opportunities: How Green Are Australia's Green Job Credentials?" *Environmental Politics* 5 (4): 607–631. https://doi.org/10.1080/09644019608414295

Crowley, Kate. 1996b. "Nature: Reinvention, Restoration or Preservation?" *Environmental Politics* 5 (2): 367–371. https://doi.org/10.1080/09644019608414274

Crowley, Kate. 1999. "Jobs and Environment: The 'Double Dividend' of Ecological Modernisation?" *International Journal of Social Economics* 26 (7/8/9): 1013–1027. https://doi.org/10.1108/03068299910245787

Crowley, Patrick. 2024. "Organizing Climate Jobs Rhode Island." In *Power Lines: Building a Labor–Climate Justice Movement,* edited by Jeff Ordower and Lindsay Zafir. The New Press.

Dewey, Scott. 1998. "Working for the Environment: Organized Labor and the Origins of Environmentalism in the United States, 1948–1970." *Environmental History* 3 (1): 45–63. https://doi.org/10.2307/3985426

Dewey, Scott. 2019. "Working-Class Environmentalism in America." In *Oxford Research Encyclopedia of American History,* edited by Scott Dewey. Oxford University Press. https://doi.org/10.1093/acrefore/9780199329175.013.690

Dryzek, John S., ed. 2003. *Green States and Social Movements: Environmentalism in the United States, United Kingdom, Germany, and Norway.* Oxford University Press.

Fellner, Kim. 1998. "Unions and Environmentalists: Will Climate Change Change the Climate?" *National Organizers Alliance Newsletter* (June): 5–8, 39–40.

Foster, David, Alex Maranville, and Sam F. Savitz. 2023. "Jobs, Emissions, and Economic Growth— What the Inflation Reduction Act Means for Working Families." Energy Futures Initiative. January.

Gadzanku, Sika, Alexandra Kramer, and Brittany Smith. 2023. "An Updated Review of the Solar PV Installation Workforce Literature." National Renewable Energy Laboratory. April 19. https://doi.org/10.2172/1971876

Gallie, W.B. 1955–1956. "Essentially Contested Concepts." *Proceedings of the Aristotelian Society* 56: 167–198.

Goodstein, Eban. 1999. *The Trade-Off Myth: Fact and Fiction about Jobs and the Environment.* Island Press.

Gordon, Robert W. 2004. "Environmental Blues: Working-Class Environmentalism and the Labor-Environmental Alliance, 1968–1985." Ph.D. dissertation. Wayne State University.

Grossman, Richard, and Gail Daneker. 1979. *Energy, Jobs and the Economy.* Alyson Publications.

Höffken, Johanna, and M.V. Ramana. 2024. "Nuclear Power and Environmental Injustice." *WIREs Energy and Environment* 13 (1): e498. https://doi.org/10.1002/wene.498

Hultgren, John. 2025. *The Smoke and the Spoils: Anti-Environmentalism and Class-Struggle in the United States.* MIT Press.

ILR Climate Jobs Institute. 2021. "NYS Passes Labor Standards on Clean Energy Work." ILR School. Cornell University. May 7. https://tinyurl.com/58f37br2

International Labour Organization. 2013. "Sustainable Development, Decent Work, and Green Jobs." April 19. https://tinyurl.com/ynmdpefs

Jones, Newton. 2019. "Climate Solutions Should (and Can) Save Our Planet and Our Jobs." The International Brotherhood of Boilermakers. September 10. https://tinyurl.com/mr3fdupc

Kazis, Richard, and Richard L. Grossman. 1982. *Fear at Work: Job Blackmail, Labor, and the Environment.* Pilgrim Press.

Kouri, Rosa, and Amelia Clarke. 2014. "Framing 'Green Jobs' Discourse: Analysis of Popular Usage." *Sustainable Development* 22 (4): 217–230. https://doi.org/10.1002/sd.1526

Kyne, Dean, and Bob Bolin. 2016. "Emerging Environmental Justice Issues in Nuclear Power and Radioactive Contamination." *International Journal of Environmental Research and Public Health* 13 (7): 700. https://doi.org/10.3390/jerph1

Labor Network for Sustainability. 2009. "Labor's Changing Approach to Climate Change." April 30. https://tinyurl.com/3bwmfzb9

Leopold, Les. 2007. *The Man Who Hated Work and Loved Labor.* Chelsea Green Publishing.

Little, Amanda. 2005. "Apollo Alliance Now Shooting for the Statehouse Rather Than for the Moon." *Grist.* September 29. https://grist.org/climate-in/apollo1/

Loomis, Erik, ed. 2015. "Working-Class Forests." In *Empire of Timber: Labor Unions and the Pacific Northwest Forests*, 89–122. Studies in Environment and History. Cambridge University Press. https://doi.org/10.1017/CBO9781

Loomis, Erik. 2022. "Saylesville Commemoration Speech by Erik Loomis." Rhode Island Labor History Society. https://tinyurl.com/y5y8h8dr

Loomis, Erik. 2018. "The Day in Labor History: May 23, 1950." *Lawyers, Guns & Money.* https://tinyurl.com/2kj8jabu

Loomis, Erik. 2021. "Working-Class Environmentalism: The Case of Northwest Timber Workers." In *The Palgrave Handbook of Environmental Labour Studies*, edited by Nora Räthzel, Dimitris Stevis, and David L. Uzzell. Palgrave Macmillan. https://doi.o/10.1007/978-3-030-71909-8

Luce, Stephanie. 2022. "Another World (of Work) Is Possible." In *The Green New Deal and the Future of Work*, edited by Craig J. Calhoun and Benjamin Y. Fong, 53–77. Columbia University Press.

Maher, Neil M. 2009. *Nature's New Deal: The Civilian Conservation Corps and the Roots of the American Environmental Movement.* Oxford University Press.

Malm, Andreas. 2016. *Fossil Capital: The Rise of Steam Power and the Roots of Global Warming.* Verso.

Manzo, Frank, Andrew Wilson, and Russell Kashian. 2022. "Building Good Local Jobs on Utility-Scale Clean Energy Projects in Wisconsin: The Impact of High-Road Labor and Contracting Standards." Midwest Economic Policy Institute. March 22. https://tinyurl.com/y7pv42zj

Mattera, Philip. 2009. "High Road or Low Road? Job Quality in the New Green Economy." Summary Report. Good Jobs First. https://tinyurl.com/46az8t53

Mayers, Matthew. 2024. "Organizing a Just Transition." In *Power Lines: Building a Labor-Climate Justice Movement*, edited by Jeff Ordower and Lindsay Zafir. The New Press.

Meadowcroft, James. 2009. "What About the Politics? Sustainable Development, Transition Management, and Long-Term Energy Transitions." *Policy Sciences* 42 (4): 323–430. https://tinyurl.com/53eb4d3f

Miller, Alan S. 1980. "Towards an Environmental/Labor Coalition." *Environment* 22 (5): 32–39. https://doi.org/10.1080/00139157.1980.9929779

Mitchell, Timothy. 2011. *Carbon Democracy: Political Power in the Age of Oil.* Verso.

Monneret, Emma. 2020. "Ignacy Sachs and Ecodevelopment: The Concept and Its Posterity." M.A. thesis. Université Lumière Lyon 2. https://tinyurl.com/szurt9kf

Montrie, Chad. 2008. *Making a Living: Work and Nature in the United States.* University of North Carolina Press.

Montrie, Chad. 2018. *The Myth of Silent Spring: Rethinking the Origins of American Environmentalism.* University of California Press.

National Centers for Environmental Information. 2024a. "Costliest Tropic US Cyclones." Government Data Set. https://tinyurl.com/45fvebdw

National Centers for Environmental Information. 2024b. "U.S. Billion-Dollar Weather and Climate Disasters." https://tinyurl.com/yc6m7ews

National Labor Relations Board. 1971. Case 23-CA-3348. July 8.

Neale, Jonathan. 2008. *Stop Global Warming: Change the World.* Bookmarks Publications.

Neale, Jonathan. 2014. *One Million Climate Jobs: Tackling the Environmental and Economic Crisis.* Third edition. Campaign Against Climate Change. https://tinyurl.com/8muykwb7

Nicholson, Phillip. 2004 *Labor's Story in the United States.* Temple University Press.

Ocasio-Cortez, Alexandria. 2019. "Recognizing the Duty of the Federal Government to Create a Green New Deal." Pub. L. No. 109. https://tinyurl.com/2v93ccsw

Ordower, Jeff. 2024. "Our Work Is What Makes the Food System Go." In *Power Lines: Building a Labor-Climate Justice Movement,* edited by Jeff Ordower and Lindsay Zafir. The New Press.

Osaka, Shannon. 2020. "Obama's Recovery Act Breathed Life into Renewables. Now They Need Rescuing." *Grist.* June 1. https://tinyurl.com/2p7bc5rw

Page, Joseph, and Mary-Win O'Brien, eds. 1973. *Bitter Wages: Ralph Nader's Study Group Report on Disease and Injury on the Job.* Penguin Books.

Pollin, Robert. 2009. "Doing the Recovery Right." *The Nation.* February 16.

Pollin, Robert, Heidi Garrett-Peltier, James Heintz, and Helen Scharber. 2008. "Green Recovery: A Program to Create Good Jobs and Start Building a Low-Carbon Economy." Center for American Progress. https://tinyurl.com/4puvyvbz

Radkau, Joachim. 2014. *The Age of Ecology: A Global History.* Polity Press.

Rector, Josiah. 2018. "The Spirit of Black Lake: Full Employment, Civil Rights, and the Forgotten Early History of Environmental Justice." *Modern American History* 1 (1): 45–66.

Renner, M.G. 1991. "Jobs in a Sustainable Economy." Worldwatch Institute.

Renner, Michael. 2000. "Working for the Environment: A Growing Source of Jobs." Worldwatch Institute.

Renner, Michael, Sean Sweeney, and Jill Kubit. 2008. "Green Jobs: Working for People and the Environment." Worldwatch Institute. https://tinyurl.com/bdh4dksf

Roberts, Cecil, and Lonnie Stephenson. 2019. "AFL-CIO Letter to Sen. Ed Markey and Rep. Ocasio-Cortez," March 8. https://tinyurl.com/547p3x43

Sanders, Jeffrey C. 2021. *Razing Kids: Youth, Environment, and the Postwar American West.* Cambridge University Press.

Savage, Larry, and Dennis Soron. 2011. "Organized Labor, Nuclear Power, and Environmental Justice: A Comparative Analysis of the Canadian and U.S. Labor Movements." *Labor Studies Journal* 36 (1): 37–57. https://doi.org/10.1177/0160449X10389746

Saul, Josh. 2023. "Goldman Sees Biden's Clean-Energy Law Costing US $1.2 Trillion." *Bloomberg.* March 23. https://tinyurl.com/mpaa6ewm

Silver, Beverly. 2003. *Forces of Labor: Workers' Movements and Globalization Since 1870.* Cambridge University Press. https://doi.org/10.1017/CBO9780511615702

Smith, Adrian. 2014. "The Lucas Plan: What Can It Tell Us About Democratising Technology Today?" *The Guardian.* January 22. https://tinyurl.com/44d4msd4

Stanef-Puică, Mihaela-Roberta, Liana Badea, George-Laurentiu Șerban-Oprescu, Anca-Teodora Șerban-Oprescu, Laurentiu-Gabriel Frâncu, and Alina Crețu. 2022. "Green Jobs—A Literature

Review." *International Journal of Environmental Research and Public Health* 19 (13): 7998. https://doi.org/10.3390/walk

Stevis, Dimitris. 2013. "Green Jobs? Good Jobs? Just Jobs?: US Labour Unions Confront Climate Change." In *Trade Unions in the Green Economy: Working for the Environment*, 179–195. Routledge.

Stevis, Dimitris. 2014. "US Labour Unions and Climate Change: Technological Innovations and Institutional Influences." In *Climate Innovation: Liberal Capitalism and Climate Change*, edited by Neil E. Harrison and John Mikler, 164–188. Palgrave Macmillan.

Stevis, Dimitris. 2019. "Labour Unions and Green Transitions in the USA: Contestations and Explanations." Working Paper. York University. Adapting Canadian Work and Workplaces to Respond to Climate Change. https://tinyurl.com/46hz5d55

Stevis, Dimitris. 2021. "Embedding Just Transition in the USA: The Long Ambivalence." In *Handbook of Environmental Labour Studies*, edited by Nora Räthzel, Dimitris Stevis, and David Uzzell, 591–619. Palgrave Macmillan.

Stevis, Dimitris. 2023. *Just Transitions: Promise and Contestations*. Cambridge University Press.

Stewart, Andrew. 2018. "Sorry Democrats, the Green Party Came Up with the Green New Deal!" *Counterpunch*. November 29. https://tinyurl.com/5ad45rh4

Sweeney, Sean. 2012. "Resist, Reclaim, Restructure: Unions and the Struggle for Energy Democracy." ILR School. Cornell University. Trade Unions for Energy Democracy. https://tinyurl.com/yc2xmpt2

Treuer, David. 2021. "Return the National Parks to the Tribes." *The Atlantic*. May. https://tinyurl.com/y2faty4v

Uehlein, Joe. 2010. "Opinion: Earth Day, Labor, and Me." *Common Dreams*. September 1. https://tinyurl.com/4nfs6pxc

United Nations Environment Programme. 2008. "Green Jobs: Towards Decent Work in a Sustainable, Low-Carbon World." https://tinyurl.com/3c3jywy5

United Steel Workers. 1990. "Our Children's World." https://tinyurl.com/mr36kyss

U.S. Bureau of Labor Statistics. 2011. "Green Jobs." https://www.bls.gov/green/home.htm

U.S. Department of Energy. 2021. "Justice40 Initiative." https://tinyurl.com/yfzcp9fv

U.S. Department of Labor. 2023. "Prevailing Wage and the Inflation Reduction Act." https://www.dol.gov/agencies/whd/IRA

U.S. Energy Information Administration. 2020. "Annual Energy Outlook 2020." January. https://tinyurl.com/yu7rt3fa

Vachon, Todd E. 2023. *Clean Air and Good Jobs: U.S. Labor and the Struggle for Climate Justice*. Temple University Press.

Velut, Jean-Baptiste. 2011. "A Brief History of the Relations Between the U.S. Labor and Environmental Movements (1965–2010)." *Revue Française d'études Américaines* 129 (3): 59–72. https://doi.org/10.3917/rfea.129.0059.

Vig, Norman J., and Michael Kraft, eds. 2013. *Environmental Policy: New Directions for the Twenty-First Century*. Eighth edition. Sage.

White House. 2023. "Biden–Harris Administration Launches American Climate Corps to Train Young People in Clean Energy, Conservation, and Climate Resilience Skills, Create Good-Paying Jobs and Tackle the Climate Crisis." September 20. https://tinyurl.com/4vece85e

Woodcock, Leonard. 1972. "Labor and the Economic Impact of Environmental Control Requirements." Institute of Industrial Relations, University of California, Berkeley.

World Commission on Environment and Sustainable Development. 1987. "Our Common Future."
https://tinyurl.com/24zdz8m8

Young, Jeff. 2021. "Mine Workers' Leader Wants to Save Last Coal Jobs as Biden Tackles Climate."
United Mine Workers of America. April 20. https://tinyurl.com/2av9dcee

# Working Conditions in the U.S. Solar Industry: Findings and Learnings from Studies in New York and Texas

JILLIAN MORLEY
AVALON HOEK SPAANS
*Cornell ILR Climate Jobs Institute*

## ABSTRACT

We discuss the lack of research available on the conditions and demographics of the solar workforce, as well as the importance of conducting surveys that directly reach these workers. Despite the solar industry's much-needed and rapid growth, the existing research on its laborers' working conditions is limited. After supporting two key surveys—in New York State (n = 264) and Texas (n = 842), which provided the first research of its kind directly with solar workers—we provide a detailed overview of the findings and lessons learned based on comparing each study. Both surveys of nonunion solar workers uncovered evidence showing significant differences in working conditions for workers of color. The findings from Texas' sample indicate that there may be racial and language disparities in health and safety on solar worksites. In New York's sample, over 40% of workers had more than one employer for their solar work, suggesting that additional data must be collected from workers directly to supplement employer-collected datasets.

## CENTERING WORKERS IN THE CLIMATE TRANSITION

Solar power is the fastest-growing renewable energy source in the United States, ushering in the new "green" economy, which requires a workforce of scale (Al Mubarak, Rezaee, and Wood 2024, Tabassum et al. 2021). A recent (2023) study by Curtis and Marinescu found that online postings for solar jobs have more than tripled since 2010, and this growth has also spurred a substantial amount of investigation into workforce development and expansion approaches in the solar industry (Lesser 2023; Moe et al. 2024; Srivastava, Ayala, Tuttle, and Moe 2024). This research is coming at a crucial point in the economic transition, as recent modeling indicates that the number of workers from inevitably declining, carbon-intensive fields (like petroleum refining or coal and other mining industries) that are moving into renewable energy jobs like solar appears to be remarkably low, at less than 1% (Curtis, O'Kane, and Park 2024). Beyond workforce age, one of the hypothesized reasons for this low rate of transition is the issue of job quality; however, the literature examining workers' experiences on solar worksites is severely limited (Gadzanku, Kramer, and Smith 2023). This likely stems from the fact that data on the solar workforce are primarily sourced from employers, such as the National Solar Jobs Census (Interstate Renewable Energy Council 2024) and the Bureau of Labor Statistics' (BLS) Occupational Employment and Wages survey (U.S. Bureau of Labor Statistics 2024a), rather than workers on solar sites themselves. To ensure that the urgently needed transition to a renewable economy is both possible and equitable, research on the quality of jobs that centers the worker perspective in these industries must catch up to the speed at which such jobs are growing. This chapter provides readers with preliminary guidance on how to approach research on solar workers based on essons learned through initial surveys of the population.

In an attempt to forge a pathway forward for filling the literature gap on working conditions in the solar industry, we supported the development of surveys for solar workers in the states of New York (Hoek Spaans and Morley 2024) and Texas (Texas Climate Jobs Project, Climate Jobs Institute, and Organized Power in Numbers 2024). Because of the unknowns of the true statewide populations of solar workers, the transience of this workforce, workforce growth, and the lack of union respondents, neither of the samples collected can be considered representative. Yet these two studies provide a novel first look at the employment structures and potential issues in the solar industry and provide a baseline upon which to refine existing data collection processes and develop new approaches to understanding the solar workforce. For example, with over 40% of the New York sample reporting that they had more than one employer, and large portions of both samples reporting substantial solar work–related travel within and across states, responses from both surveys confirmed the hypothesis that additional research on cross-state employment and contracting mechanisms within the solar industry must be done to form accurate measures of workforce growth and demographics. Also confirmed by these survey responses was the hypothesis that evaluating worker-side data is necessary to understand the true prevalence of labor violations and other legal considerations within this industry. In New York, responses indicate that possible incidents of unreported wage theft and injury may have occurred on solar worksites during the study period, and in Texas, similar experiences were observed, with particularly concerning data around heat-related illness and breaks. Another issue with potential legal implications was the surprisingly low rate of self-identified apprentices (0.7%) across both the solar and wind surveys in Texas, which is far lower than the Inflation Reduction Act requirement that 10% of total job hours must be performed by apprentices on projects started before 2023 in order for a solar project to be eligible for the act's investment tax credit (Internal Revenue Service 2025).

These studies also bring into bare light onto additional questions about disparities, discrimination, and related legal considerations. Across both studies, racial, ethnic, and language-based disparities were observed in a variety of categories, including wages, nonwage benefits, job tenure, and experiences with occupational injuries. Racial and ethnic disparities were even observed regarding which workers witnessed fatalities on their solar sites. Although the share of female and nonmale solar installers was low for both the New York and Texas samples, for the Texas sample, there were enough female installers surveyed to identify a $2,700 annual gap in wages. Furthermore, interview-based evidence from Texas also indicated that sexual harassment and retaliatory layoffs could be an issue for women in this industry. These observed disparities emphasize the importance of ensuring that demographic estimates for the solar workforce, if based on employer-side data, are also developed alongside worker-side data to maximize accuracy, as race/ethnicity and gender-based inequality within the solar industry must be a key topic of focus for future research. The issue of specificity related to potential incidents of unreported discrimination is also essential for researchers and policy makers to understand, and collection of this type of information at such a granular level is only possible through worker-side data collection, particularly through qualitative accounts of experiences.

Understanding the working conditions faced by solar installers and maintenance workers is important for researchers and policy makers and also presents an opportunity for workers themselves to find solidarity amid the challenges they and others face and take action toward self-advocacy. For example, in the New York study, through an assessment of New York State Energy Research and Development Authority data, the top distributed-scale solar development companies were identified for each state region (Hoek Spaans and Morley 2024). In many cases, the top developers in each region had completed thousands of solar projects in the two-year study period, meaning that many projects were likely built concurrently. With different individuals on each worksite, similarities in working conditions may not be immediately apparent without relevant research. While lack of awareness of shared challenges across the workforce may present an obstacle for worker organizing, the buildout of solar being led by large national corporations may actually enable organizing efforts once shared challenges have been illuminated.

## LACK OF EMPIRICAL STUDIES ON WORKING CONDITIONS IN THE SOLAR INDUSTRY
### Background and Gaps in the Literature

In 2023, Gadzanku, Kramer, and Smith conducted a review of the existing literature on the solar workforce using Google Scholar, Scopus, and NREL library resources to conduct searches, and added additional research based on reviewer recommendations. Although the initial scope of the review was limited to academic and peer-reviewed literature specifically focused on solar installation jobs, the limited research available necessitated a widening of this scope. This resulted in the inclusion of literature on similar industries, such as construction, and alternative sources, such as technical reports or media reports. One of the main topics examined in this review was workforce well-being, and across all the literature identified, Gadzanku, Kramer, and Smith (2023) identified 12 major gaps (shown in Table 1) in the available literature on the solar workforce at the time the report was written. When literature was compiled for this chapter, the literature review yielded direct citations from only nine distinct works on Google Scholar, ranging from adjacent research on the electric vehicle manufacturing workforce (Combemale et al. 2024) and the role of global labor migration on meeting increased workforce demand (Huckstep and Dempster 2024) to the potential impacts of high-road labor standards on solar and related projects (Appel and Hammerling 2023; Boettner 2024). Across all works citing the review, most discussion overlapping with the literature gaps identified by Gadzanku, Kramer, and Smith (2023) was incidental and was often theoretical in nature—for example, Appel and Hammerling's 2023 discussion of how misclassification has been addressed through legislation in other sectors. Although each of these reports provided valuable insight into the clean energy economy and solar industry, none was able to report on experiences from the workers themselves.

While much more research is necessary to begin filling these literature gaps, the Texas and New York studies have offered some preliminary findings that are relevant to answering the questions posed in Gadzanku, Kramer, and Smith's review. More importantly, these studies have provided valuable insights into best practices for studying the solar workforce in future research.

Looking beyond Gadzanku, Kramer, and Smith's 2023 review of the solar workforce literature, we have also drawn on articles identified through iterative and ongoing systematized review processes. For the sake of brevity, the full scope of work identified on these topics will not be included in this chapter; however, in order to further illustrate the presence of a gap in the literature on solar working conditions, an abbreviated review process is outlined below for replicability, with an example of the findings from one of the search terms applied.

The lack of literature on solar working conditions can be exemplified through a brief discussion of the results yielded using the search term "solar workforce." Given the minimal academic literature available, research was identified through keyword searches in Google Scholar and then sorted for relevance to capture the widest range of article types. If the search yielded relevant articles on the first page of the results, the first 60 articles were evaluated for suitability. Articles were saved for additional evaluation if they fell within the defined scope (U.S.-based or nonregional research, workforce-focused topics, and publication after Gadzanku, Kramer, and Smith's review was made publicly available from May of 2023 and up to December of 2024). Of the 60 articles initially evaluated, 28 fell within the defined scope. However, 14 of those articles were focused solely on training programs, four examined workforce growth and/or the need for workforce growth, and three (including one of our studies) had already been identified in the works that had cited Gadzanku, Kramer, and Smith's 2023 review.

Table 1. Progress Made on Literature Gaps

| Gaps identified by Gadzanku, Kramer, and Smith 2023 (pp. 34–35) | Preliminary data from New York study | Preliminary data from Texas study | Any discussion of these gaps in articles citing Gadzanku, Kramer, and Smith |
|---|---|---|---|
| "Safe and healthy working conditions specifically for solar workers, separate from general construction and electrical codes" | Yes, this study provides worker reported data on injuries vs. formal reporting, as well as subjective feelings of safety. | Yes, this study provides several findings on health and safety issues, as well as potential risk factors. | Appel and Hammerling 2023; Boettner 2024 |
| "Standardized data on worker classification (and misclassification), and its intersection with solar workplace safety" | Yes, this study contains findings around multi-employer and transient, multiregional work that lacks benefits access that are relevant to worker misclassification and worker safety. | Yes, this study contains findings around transience, multiregional work, and lack of benefits that relate to worker misclassification and worker safety. | Appel and Hammerling 2023 |
| "Potential pathways, industry interest, and requirements for implementing national solar labor standards or national certifications for solar installation" | Yes, this study highlights areas where standards should be improved and potential challenges related to implementation. | Yes, this study highlights areas where standards should be improved and potential challenges related to implementation. | Appel and Hammerling 2023; Boettner 2024 |
| "Statistical analysis of whether minorities and women have fair access to construction opportunities on a trade-by-trade basis" | Yes, this study breaks down several self-reported experiences by race and ethnicity for solar construction workers. | Yes, this study breaks down several self-reported experiences by race and ethnicity for solar construction workers. | None identified in this search |
| "Contextualizing the intersection between energy justice, workforce discrimination, and well-being, especially regarding policies like prevailing wage." | Yes, this study captures information at the intersection of energy justice, workforce discrimination, and well-being. | Yes, this study captures information at the intersection of energy justice, workforce discrimination, and well-being. | Appel and Hammerling 2023; Boettner 2024 |
| "Best practices for recruitment and training efforts for solar PV installation jobs" | Yes, this study highlights working conditions and documents how workers receive training, which could be barriers to recruitment. | Yes, this study highlights working conditions, and lack of apprentices on jobsites, which could be barriers to recruitment. | (Appel and Hammerling 2023; Boettner 2024; Dutta et al. 2024) |

(Table 1 continues next page)

## Table 1. Continued

| | | | |
|---|---|---|---|
| "Training and workforce opportunities for disenfranchised workers, like those recovering from substance abuse or formerly incarcerated individuals " | Yes, this study found that substance use may be an issue on jobsites, indicating potential challenges for these populations. | No, knowledge gaps remain on training and workforce opportunities for disenfranchised solar workers in Texas. | None identified in this search |
| "Labor requirements and conditions related to solar manufacturing and the solar supply chain, both globally and within the United States" | No, knowledge gaps remain on working conditions in localized solar manufacturing supply chains in New York State and the United States. | Yes, limited information was captured on clean energy manufacturing workers in Texas. Knowledge gaps are persistent in manufacturing specific to the solar industry. | Combemale et al. 2024; Dutta et al. 2024; Huckstep and Dempster 2024 |
| "Labor impacts and indirect jobs related to increases in domestic solar manufacturing activities" | No, knowledge gaps remain on job creation impacts of domestic solar manufacturing activities in New York and the United States. | Yes, limited information was captured on clean energy manufacturing workers in Texas. Knowledge gaps persist in increasing domestic manufacturing specific to the solar industry. | Combemale et al. 2024; Huckstep and Dempster 2024 |
| "Improved resolution for NAICS codes with data relevant to the solar industry and solar-specific workers" | No, knowledge gaps remain on improved resolution for NAICS codes. | No, knowledge gaps remain on improved resolution for NAICS codes. | None identified in this search |
| "Environmentally, economically, and socially responsible mining projects and effective recycling initiatives for strategic materials necessary for solar PV projects, including domestic materials production and innovation" | No, knowledge gaps remain on responsible material use in the solar industry in New York and the United States. | No, knowledge gaps remain on responsible material use in the solar industry in Texas and the United States. | Dutta et al. 2024 |
| "Workforce opportunities for Native American or tribal communities" | No, knowledge gaps remain in solar workforce opportunities in tribal communities in New York but this study demonstrates the need for additional research on this community's experiences. | No, knowledge gaps remain in solar workforce opportunities in tribal communities in Texas but this study demonstrates the need for additional research on this community's experiences. | None identified in this search |

Of the remaining works identified through this search, three included original research. The first article, by Mulvaney (2024), applied an energy justice theoretical framework to the growth of solar power and discussed the potential for unjust labor dynamics at every stage of solar power production. The second, a conference paper by Behrani et al. (2023), provided a review of occupational and safety risks in solar work. Many of the risks identified by Behrani et al. (2023) had already been outlined in previous research, such as Duroha and Macht's (2023) systematic review, including falls, heat stress, electrical hazards, etc., but Behrani et al. also provided some discussion of the lack of research on psychosocial hazards in solar work. The final article, by Solis, Oden, Lieberknecht, and Liu (2025), contributed novel data analysis utilizing a mixed-methods approach. Solis, Oden, Lieberknecht, and Liu first used data from the BLS and the Department of Labor's O*NET initiative to develop "green" workforce development and job quality projections and described these results alongside additional content analysis based on interviews with 13 representatives from workforce development organizations and 12 city officials. They concluded that "green" jobs have the potential to provide quality career opportunities and that local governments can play an important role in shaping the economic development that results from climate change mitigation.

Literature identified through other searches and methods was also continuously evaluated for its applicability to research on solar working conditions in domestic contexts and can be seen throughout this chapter. Although the body of literature on solar industry workers appears to be growing, there is still a need for additional research on the quality of the jobs being created by the rapid buildout of solar power and renewable energy in the United States. Throughout the existing research, there is often discussion around the necessity of, or approaches to, ensuring that green jobs are high-quality jobs that do not reinforce existing inequalities; however, there is a limited understanding of workers' experiences.

These topical gaps in the solar industry workforce literature likely stem from one core issue: the lack of generalizable data sourced directly from workers on their experiences in the solar industry. Our studies sought to explore which recruitment methods work best when examining the solar workforce population and gain a basic understanding of areas to build on in future research. Although the samples collected for these studies cannot be considered representative of the solar workforce at large, they establish an important first look at the areas in which existing data may not provide a full picture of the solar industry at either the state or national levels.

In their review of existing literature, Gadzanku, Kramer, and Smith (2023) shared demographic estimates based on available BLS and IREC data, but highlighted the fact that, within the 2022 BLS data from their Occupational Employment and Wages survey, the category of "Solar Photovoltaic Installers" was estimated to capture only about 10% of the total 2021 solar workforce based on the data collected by IREC for the National Solar Jobs Census in 2022, as it is likely that solar work is also done by laborers in other occupation categories. While both of these data sources provide essential insight into the state of the solar industry and its workforce, both sources rely on surveys of employers, which limits the information that can be collected on workers' experiences. In particular, data sourced from surveys of employers could include response bias around topics related to workers' rights violations, such as racial discrimination, wage theft, unreported safety issues, and more, resulting in a lack of information on the true prevalence of such practices. To mitigate potential response bias present in employer-side data, research based on robust worker-side data must be available.

The lack of data available on workers' rights violations is particularly relevant for research on the solar photovoltaic (PV) industry when considering the contracting mechanisms common in the solar industry and in construction more generally (Davidson 2023) and the implications that such nonstandard arrangements have for worker well-being (Gadzanku, Kramer, and Smith 2023). Although Gadzanku, Kramer, and Smith found little academic research on nonstandard contracting mechanisms or their impact on workers within the solar-specific context, they identified relevant journalistic reporting that highlighted the negative experiences of solar workers, particularly out-of-state workers hired through temp agencies. The anecdotal evidence available through the investigative reports included in their review indicates that additional research should be done on safety and nonwage compensation for solar workers hired in nonstandard arrangements.

Furthermore, the adjacent construction sector has historically been highly dependent on multi-employer work (Weill 2005). If there is substantial overlap between these sectors, exclusive dependence on employer-side data may lead to issues in accurately understanding the industry. Reporting on wages, demographics, and job creation that is based on employer-side data may also provide an inadequate perspective, as an employer's knowledge about each employee's full working arrangements may be limited. For example, it is possible that solar workers with more than one employer—over 40% of the New York State sample—could also be represented in multiple employers' surveys. The impact of such occurrences on job, wage, and demographic estimates is currently unknown.

## Experimental Outreach Methods to Reach Solar Workers on the Ground

To make progress toward filling gaps in the collected data and literature, we supported two exploratory studies of solar workers. The first study, which surveyed solar installation and maintenance workers on projects in New York State, partially shaped the design and administration of the second study, which examined solar workers on projects in Texas. Both study protocols were submitted to the Cornell University Institutional Review Board (IRB) for determination, and the New York State–based study protocol IRB0145089 was granted an exemption from full review. The Texas study's data (protocol IRB0146572) was collected by an external organization and only provided to Cornell researchers in a deidentified form, so the project was not designated as human subjects research by the IRB and did not require approval. In both the New York and Texas studies, the total state population of solar installers or solar construction workers was treated as unknown. Because of this, both studies were intended to be exploratory, mixed-methods investigations into the most effective strategies for reaching workers on the ground. The overarching goal of these projects was to support the design of future studies by creating a foundation upon which to iterate.

## Exploring the Conditions of the New York Solar Workforce

In the "Exploring the Conditions of the New York Solar Workforce" study, the target population was solar installation and maintenance workers over the age of 18 who had worked on at least one solar project in the state between December 2021 and September 2023. The survey was conducted virtually in partnership with research firm Social Science Research Solutions (SSRS) from December 2022 to September 2023 and focused on utilizing outreach through respected solar workforce groups, such as training centers, job boards, academic partners, and nonprofit community organizations. Eight institutions distributed the study to their membership through posts on private online workforce group pages, email invitations by solar workforce trainers, and virtual flyers. It was hypothesized that successful outreach to an unknown population of workers could occur through the distribution of organizations trusted by these workers.

It was also assumed that worker landscapes would look different across each state as workers are trained, hired, and entered the industry through varying community organization, job training, and industry organizations. In New York, a landscape analysis was conducted with the intention of understanding how solar installation and maintenance workers from all scales of solar installation (utility, distributed, and community solar) were placed on projects and entered the industry within New York State. This analysis was conducted during the design phase of the study through interviews with experts in the field, such as union, industry, and workforce leaders, as well as desk research focused on community organizations, unions, industry groups, and community colleges that had programs with a pipeline for workers to enter the solar industry. Not all groups that were identified agreed to distribute the survey, with only eight organizations distributing the final survey. One of the most successful forms of outreach came from the distribution of a digital flyer by a solar worker on "American Solar Farms," a closed Facebook group and national solar job board, designed by and for solar workers to find solar jobsites. This distribution is believed to be successful because of direct partnership with workers who are a part of the coalition Green Workers Alliance.

The survey utilized a snowball sampling method with a financial incentive of $20 upon completion of the survey and $10 per survey referral to other solar workers. The snowball design of the study was created, as solar installers and maintenance workers were assumed to work on crews and could directly distribute the survey to other solar workers on their jobsites. Unique links were developed for all surveys that were distributed, and sharing of links was tracked and monitored. It was found that the solar survey had additional unplanned digital sharing, where the survey was shared by individuals and other organizations via social media and other methods of communication. Strict data quality measures were implemented to ensure any unintentional sharing of links or the survey did not affect the quality of the study. The Climate Jobs Institute worked with SSRS to filter out nonvalidated responses and protect against potential bot completions.

The questionnaire was designed with feedback from industry, workforce, and academic partners and was devised to capture working conditions on New York State solar jobsites, including but not limited to, pay, safety, and discrimination. Key demographic information was also captured, including race, age, gender, and location of residence. A notable result of the New York study (that will be discussed further and led to shifts in the design of the Texas study) were the results from the study that revealed the respondents sampled were transient, with a proportion not permanently residing in New York State, as well as having multiple employers. Another unexpected feature of the New York sample was the lack of union respondents. Despite collaboration with union partners, 263 of the 264 (99.6%) respondents reported that they did not belong to a union. Future research must identify the challenges related to surveying the unionized solar workforce to better understand how union membership impacts job quality. One hypothesis is that union solar workers may have been less interested in the $20 monetary incentive of this survey.

The data analysis, like its collection and validation, was done in partnership with SSRS. Much of what is discussed in the findings for this chapter is based on descriptive statistics, but deeper analysis is described in the original report. The sample size was somewhat limiting in regard to what testing was possible, but after an initial round of data exploration and significance testing by the Cornell team, we shared a series of hypotheses and tests we were interested in seeing the results of with the team at SSRS. SSRS was able to run four linear regression models in a stepwise manner, looking at how outcomes differed based on basic demographic information, worksite type, certifications, and regions worked.

Although one intended goal of the New York study was to collect qualitative data as a complement to the survey, the number of participants who were willing to be contacted for semi-structured interviews after completion of the survey was low. Ultimately, only one participant chose to complete an in-depth interview after completing the survey. The reimbursement rate for those who completed interviews was $100, and this amount may need to be assessed further compared to market value. However, given the more successful recruitment of interview participants in Texas, it is also possible that the lower in-depth interview participation in the New York study may have been related to data collection methods rather than compensation.

## Power and People: Working Conditions in the Texas Clean Energy Transition

"Power and People: Working Conditions in the Texas Clean Energy Transition" was designed to answer the same question: Who are the workers building out clean energy in Texas, and what are the conditions under which they work? This study was designed in collaboration with the Texas Climate Jobs Project, as well as Organized

Power in Numbers, and aimed to capture the experience of workers in Texas across clean energy manufacturing and utility-scale wind and utility-scale solar projects, with this chapter's focus being utility-scale solar workers. Based on learnings from the New York study and shifting objectives of information gathering, the solar designation of the Texas clean energy study focused on capturing information from nonunion utility-scale solar construction workers over the age of 18 who completed work in Texas from May 2023 through March 2024. The questionnaire was mirrored from the New York State solar study but was shortened based on the prevalence of incompletes in the New York State study. Additionally, the survey incentive was increased from $20 to $50 to increase the number of survey completions.

The Texas study relied on two main forms of sampling: digital outreach led by Organized Power in Numbers and field outreach led by the Texas Climate Jobs Project. Digital outreach consisted of multiple methods, including peer-to-peer texting from list building, social media ads, and geofencing. Peer-to-peer texting was conducted in English and Spanish from commercial and public lists of potential clean energy workers in the state. All surveys were administered via single-use survey links to minimize the rate of duplicative data from individual respondents. Public information was acquired through the Freedom of Information Act, specifically focusing on building lists from community colleges and those who underwent state trade certification. Additional predictive modeling was used for list building for the targeted social media ads. Geofencing was used as a preliminary method to contact workers within specific areas of utility-scale projects who received targeted advertisements to complete the survey. Ultimately, this method was abandoned after the regulation of geofencing changed during the time of the study. Several data validation methods were employed to further prevent any inclusion of duplicative data or data from individuals outside of the target population. These validation methods included IP address and time stamp monitoring, review of survey responses by two members of the Organized Power in Numbers team, and when necessary, follow-up by phone for flagged surveys. Though digital outreach methods were extensive, the most successful recruitment method was field sampling. A team of researchers from the Texas Climate Jobs Project would visit establishments active in construction utility-scale solar projects. Much of the discussion shared in subsequent sections will be based on higher-level descriptive statistics. However, the original report includes deeper discussion of several linear and logistic regression models created by the Texas Climate Jobs Project team and reviewed by the Cornell team.

In addition to survey completions, 11 utility-scale solar workers agreed to in-depth interviews. Trust established through in-person recruitment methods may have encouraged more interview participation. The interview analysis was conducted by two coders, with the second round of open coding done using MAXQDA. A full thematic analysis was not performed; however, several of the axial codes are described in the findings section.

## SAMPLE DESCRIPTION

*Although these studies were designed in tandem, their results cannot be directly compared because of varying sampling methods, rewording and redesign of survey questions, and scale of solar projects targeted for surveying. Though the samples cannot be directly compared, the results of both these studies shed light on working in the solar industry at large and on the demographics of solar construction workers.*

In comparing each of these samples to other data, we recommend comparing the New York State race and gender numbers to the 2022 Geographic Profile of Employment and Unemployment dataset, which was the most recent version available at the time of that study (U.S. Bureau of Labor Statistics 2023), and we recommend comparing the Texas sample to the 2023 version of the same dataset. Future research may yield more appropriate data, but this dataset was chosen because its estimates are derived from the household-sourced Current Population Survey, as opposed to industry surveys. For both years, we recommend using the larger category of construction and extraction workers, which contains solar PV installation workers. This category was chosen because the construction and extraction workforce category may capture workers that perform solar work outside of their primary job functions and is better established than the solar workforce alone, allowing for a larger body of research to confirm its demographics. For age comparisons, the narrower category of solar PV installer category is recommended instead, as the construction workforce over the age of 55 doubled between 2003 and 2020 (McAnaw Gallagher 2022), and it is possible that the population of those working on solar installations could skew lower than those with established careers in other kinds of construction and extraction trades because of the recency of the sector's growth. For any statements regarding comparisons made below, the above datasets were used for each study.

### New York

Of the 264 respondents in the New York State study, the majority (approximately 83%) indicated that they identified either as Black/African American or White (Figure 1). Although the total population has been treated as unknown, it is likely that this study oversampled Black workers and undersampled White workers. However, this oversampling may give specific insights to conditions Black /African American workers face, and results on working conditions for these workers should be investigated further.

In 2023, the BLS estimated that approximately 10% of solar PV installers were between the ages of 16 and 19; however, this study had a minimum sampling age of 18. Over 55.3% of our sample reported being between the ages of 26 and 30, 39% between the ages of 31 and 45, less than 5% between the ages of 18 and 25, and less than 1% between the ages of 46 and 55. Future research should be done to understand the best way to conduct direct outreach to younger solar construction workers under the age of 25 and solar construction workers over the age of 55.

Figure 1. New York Sample Description by Race and Ethnicity

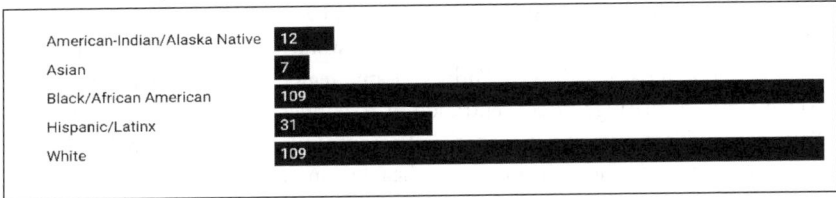

Source: Climate Jobs Institute, Cornell ILR School. Created with Datawrapper
(https://www.datawrapper.de/_/7KOOj/?v=14).

The gender distribution sampled in the New York study mirrored gender distributions of the construction and extraction workforce, with the majority of respondents reporting being male. Additionally, 1.2% of respondents self-identified as nonbinary but cannot be compared directly to distribution of the 2022 BLS annual estimate for construction and extraction workers because of the lack of nonbinary and trans classifications within the BLS dataset.

One of the most important facets of this sampled demographic is location of residence. Though the study focused on workers who had completed solar construction work in the state, not all survey respondents permanently resided in New York State (Figure 2). This opens many questions around whether local jobs are being created by the increasing demand for solar work, as the results from this study indicate some level of transience within the solar construction workforce. Though workforce transience does not inherently point to job quality issues, other results to be discussed later in this chapter indicate that transience may play a role in job quality for solar workers.

Workers within the sample reported working not only on projects within New York State but varying other states, with high concentrations of workers completing additional work in California, Texas, Florida, Illinois, Wisconsin, Michigan, Delaware, and

Figure 2. State of Current Residence, Based on New York Survey (color online).

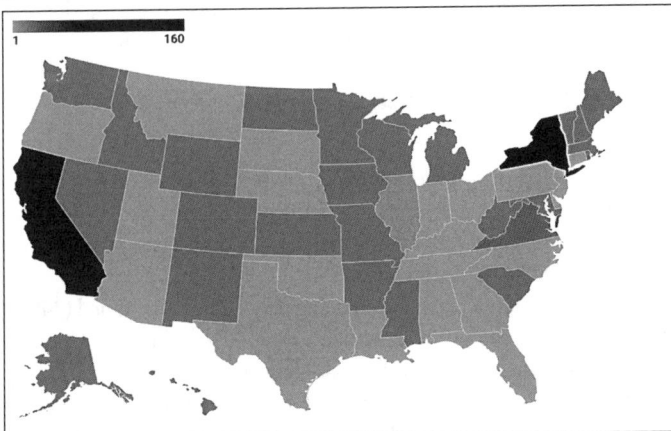

Source: Climate Jobs Institute, Cornell ILR School.
Created with Datawrapper
(https://www.datawrapper.de/_/JC7nvJ/?v=5)

Maryland. Overall, 30% of respondents reported that they had performed solar work in western states in the last year, 23.5% said they had worked in southern states or the District of Columbia, and over 20% said they had worked in the north central region.

When looking at which New York economic regions survey respondents completed work in, the highest concentration of respondents worked in western New York, central New York, and New York City. These results were compared to installation in economic regions during the applicable sample period.

## Texas

Comparing the Texas sample to the same BLS data from the following year (U.S. Bureau of Labor Statistics 2024b), there was a relatively low rate of White respondents (10.9% and a higher rate of both Hispanic/Latino respondents (55.8%), and Black respondents (23%) (Figure 3). The number of female respondents, 160, or 19% of the sample, appears to have been much higher than what would be expected for a typical estimate of construction and extraction workers, enabling a more gender-based analysis than was possible in the New York sample.

Figure 3. Composition of Texas Solar Survey Respondents by Race or Ethnicity

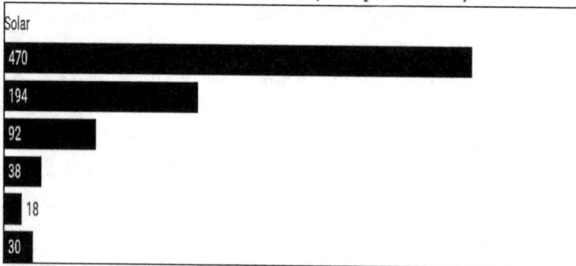

Source: Climate Jobs Institute, Cornell ILR School. Created with Datawrapper (https://www.datawrapper.de/_/qtkCu/).

Like the New York sample, many of the respondents lived in states outside of the ones where they were performing solar work (Figure 4), providing additional evidence for the argument that this may be a highly transient population.

## FINDINGS ON WORKER EXPERIENCE
### New York

Even in looking at the simplest breakdown of the New York survey's results, there are many indicators that solar installation and maintenance job quality can be greatly improved. The first example included among these indicators was the lack of benefit access across the majority (58%) of the sample.

Notable rates of respondents reported being paid by panel installed (34%) (Figure 5) or experiencing wage theft (23%), with less than 20% of those experiencing wage theft stating that the theft had been formally reported.

Similarly, 17% of the sample had either been injured or observed an injury on the jobsite, but the majority (67%) of those respondents said the injury had not been

Figure 4. Solar Survey Respondents' Permanent State of Residence,
Based on Texas Survey (color online)

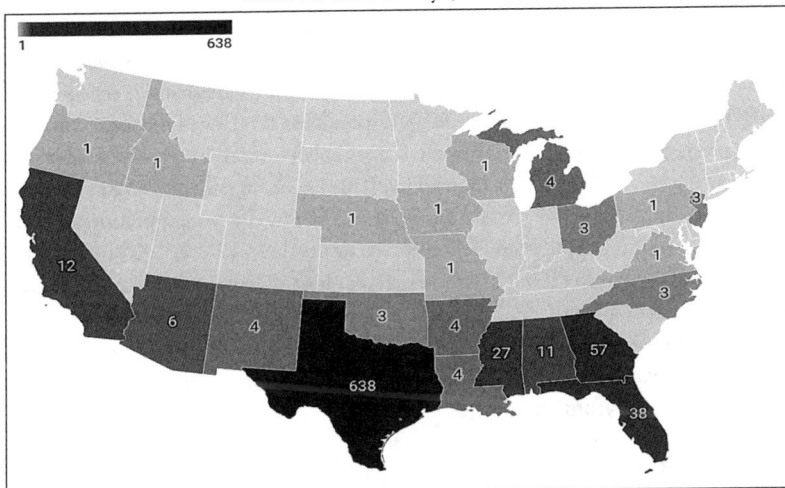

Source: Climate Jobs Institute, Cornell ILR School. Created with Datawrapper
(https://www.datawrapper.de/_/0d0UP/?v=8).

reported to OSHA, and another 22% were unsure whether it had or not. Another
example of job quality issues, among many others listed in the report, was the fact
that 52% of respondents—a rate that was associated with pay-per-panel when looking at demographics alone—agreed with the statement, "Using stimulants such as
methamphetamine to stay awake is a problem on my solar worksites in New York
State." Although this association disappeared with the inclusion of other variables,
such as regions worked or worksite type, future research should examine the role of
pay-per-panel further. Beyond the examples of job quality concerns listed in this
chapter and the original report, there were also many instances of racial and ethnic
disparities that were observed.

Figure 5. How New York Solar Workers Typically Get Paid

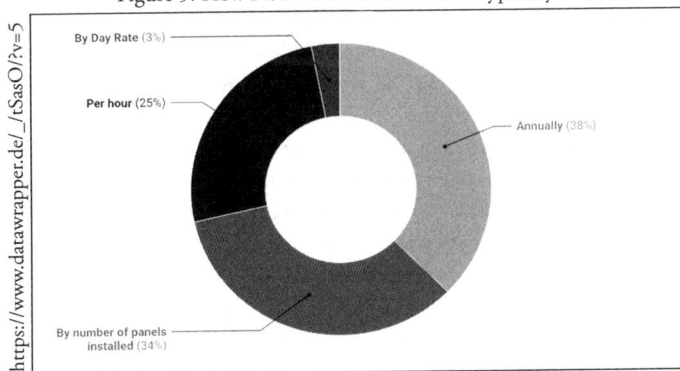

https://www.datawrapper.de/_/tSasO/?v=5

## Texas

The findings uncovered in the Texas study added additional context to the job quality concerns observed through the assessment of the New York State survey results, with a few examples outlined below. Similar to the issue of low-benefit access noted in the New York sample, many of the Texas respondents also lacked access to common nonwage benefits, and the larger sample size enabled a more meaningful look into the rates of access by type of benefit available. Of the respondents, 47% did not have access to health insurance, 72% did not have access to workers compensation or retirement benefits, and 21% even lacked access to breaks.

A substantial portion (26%) of the Texas solar respondents saw or experienced an injury on their solar worksite, and even more (48%) had personally dealt with a heat-related illness—a rate that was even higher (60%) for sample respondents without access to breaks—and that will be analyzed more thoroughly in future research. A shocking finding from this survey was that 7% of the solar survey respondents said that they had personally witnessed a fatality on their jobsite.

As with the New York study, significant racial and ethnic disparities were observed in the Texas study sample, as well as disparities related to language. Additional examples of the kinds of findings observed in both the Texas and New York studies can be found in the original report and will be elaborated on in future publications.

Another successful outcome of the Texas study's sampling approach was its success in recruitment of qualitative interview subjects. The interview findings affirmed much of what was observed in the survey data and provided more depth and context for which to build an understanding of those findings.

As shown in Table 2, insights included the substantial and disruptive amount of travel required of workers in this field, the inadequacy of certain benefits, such as per diem, and the precarious nature of the work. Simultaneously, many of the interviewed workers said they have formed strong bonds with others on their jobsites and enjoy the actual work day to day, in spite of the job quality issues observed through these studies. This final feature of the interview data highlights the fact that there is potential for these jobs to become more widely appealing careers if job quality issues are addressed, a goal that is important for facilitating the necessary scale of solar build-out.

While the strong bonds described on these solar jobsites can be seen as desirable, attention must also be paid to the way that reliance on social networks in employment can contribute to the exacerbation of labor market inequalities (Buhai and Van der Leij 2023). Disparities and inequality were observed in both the quantitative and qualitative data, and statements from the interviews about career advancement, such as the second quote in Table 2, demonstrate how social networks may act as one of the potential mechanisms for inequality. The issue of social networks and employment on these jobsites could also create protected spaces for the mistreatment of workers who fall outside of these networks. An example of gender-based mistreatment from the interviews, which was not provided in the table, was the experience of a female solar worker who described being fired a week after reporting to her supervisor that she had been experiencing sexual harassment on her jobsite. Future research

Table 2. Key Topics from Interviews with Solar Workers

| Key topic | Description | Representative quote |
|---|---|---|
| Precarity | Instability around employment | "There's no contract. You know you can get terminated at any time for whatever reason." |
| Employment and social networks | The role of social networks in recruitment and advancement | "It's not really about your hard work … I feel like most mostly these days it's about who you know" (when asked about career advancement) |
| Health/safety | Incidences related to health or safety | "As far as it being a big problem, … nothing's ever happened … most people do get a little nick or a cut, that's the way it is." |
| Positive experience with coworkers | Positive relationships born out of solar work | "My crew … they really like family, man … I don't have an issue with nobody." |
| Insufficient benefits | Insufficiency in either benefits offered or accessibility of benefits | "If we don't work on Fridays we don't get no per diem … that's a big problem … you have to have your own, provide your own housing." |
| Transience | Travel away from home for solar work | "I could go like six months without seeing my family. It's hard." |
| Enjoying work | Positive feelings about work in the solar industry | "I have a good day every day." |
| Negative experience with management | Issues with managers | "When management has their own family come in and they … they get paid more for doing less." |
| Privacy around pay | Unwillingness to discuss topics around pay | "I don't try to count nobody's money but mine." |
| Disparities/inequality | Unequal conditions/compensation between workers | "There's a lot of conflict regarding like identity, or like racial stuff" |

should examine the social connections fostered on solar jobsites and on ways to balance the resulting benefits and challenges.

## FUTURE RESEARCH: BRIDGING THE INFORMATION GAP

The data collected for these exploratory studies in New York and Texas represent a foundation for understanding working conditions and job quality in the solar industry. These studies, being the first of their kind, also exhibit many limitations. The high proportions of multi-employer work, multistate work, and variable states of residence indicate the need for supplementary, targeted sampling methods of solar installation workers beyond the traditional employer-side data collection to improve the accuracy of solar workforce estimates and the level of detail available on working conditions. Because of the high rates of multistate work and relocation observed in these samples, we caution that it is possible that the only way to ensure a sample that accurately reflects the solar workforce population would be to perform a study at the national level, inclusive of careful data validation techniques designed to prevent double counting across states. Additionally, to better understand the specifics of contracting and employment mechanisms within the solar industry, as well as industry working conditions more generally, it would be beneficial to complement survey findings with a qualitative component, such as in-depth interviews. Coordination of a national-level study of this kind would be highly resource intensive and challenging; however, the replication of similar studies on a smaller scale will contribute to an understanding of the best methods available for reaching solar installation and maintenance workers.

In addition to providing a knowledge base for new approaches to sampling the solar workforce, these studies also opened new avenues for further exploration of working conditions. One major feature missing from both studies was the ability to compare working conditions based on union membership status. Although the goal of the New York study was to reach union solar workers, the fact that only one respondent reported being a member of a union indicates a need to re-evaluate the strategy used to recruit this subgroup. Based on a wealth of previous literature on wages and unionization (Denice and Rosenfeld 2018; Parolin and VanHuevlen 2023; VanHuevlen and Brady 2022), we believe that the survey incentive may have been too low to be worthwhile to union members; however, additional research is necessary to confirm this idea. For the Texas survey, the target population was exclusively nonunion workers, so this question remains unanswered until it is addressed in future research.

Several other limitations and insights can be built on, based on the results of each study. One example that illuminates the importance of studying other subgroups more extensively is that, for all 12 of the New York respondents that identified as American Indian or Alaska Native, discrimination or harassment on solar worksites was reported, and 11 of the 12 respondents reported sexual harassment. Similarly, there were only 17 respondents that identified as female and three respondents that identified as nonbinary, limiting options for comparison of experiences by gender. In the Texas study, the findings indicated potential gender-based disparities that warrant further investigation.

In addition to limitations related to sample size, the New York study also faced challenges around question phrasing, which can be used to inform the development of following survey instruments. For many of the questions, respondents were given the option to select more than one option as they applied, and this led to additional questions about how different sample characteristics may relate to experiences. A key point that future research should address is how individuals' experiences differ across different regions or project types (i.e., residential vs. utility-scale solar). Based on preliminary analysis, there were some differences in how respondents who had worked on residential or commercial solar projects responded to questions; for example, they were less likely to report being paid per panel and likelier to report long hours, but these kinds of differences would be better suited for further examination in a qualitative study, where follow-up questions are possible, or through a quantitative study that specifically targets residential solar workers. Another issue with question phrasing observed in New York stemming from a lack of research on workers themselves was the use of terms like "crew." Response rates to questions that used the term "crew" tended to be lower, indicating that research on sociocultural or linguistic practices within the solar workforce could be useful for shaping future surveys. Finally, in comparing the results from the New York and Texas samples, it seems as though in-person over virtual recruitment methods may be more effective for studying this population, both in terms of sample size yield and interest in interview participation.

Beyond areas for improvement, the results from New York also indicate the need for thorough legal and academic investigation into the practices of individual contractors. The 34% of respondents who reported piece-rate pay, the 23% who indicated that they had experienced wage theft, and the majority (53%) of respondents who agreed that the use of stimulants like methamphetamines to stay awake on solar worksites is an issue offer valuable insight into problematic trends in the industry. The willingness to come forward about these experiences anonymously should be met with efforts to enable better enforcement of labor standards and lower-risk avenues for formal reporting of subpar conditions.

Shifting to a brief discussion of the lessons learned from the Texas study, many of the results reflected those of the New York study but on a larger scale. This included additional evidence that a large proportion of solar workers may travel between states for their work, providing further support for concerns regarding the true size and demographic makeup of the solar workforce. The additional confirmation of this finding also highlights a limitation of both samples; post-stratification demographic weighting should not be applied to either the Texas or New York dataset. The Texas results reinforce the New York findings, with further evidence of racial disparities in solar working conditions and compensation.

Beyond providing support for the initial findings from New York, the Texas study's larger sample size also allowed for the preliminary analysis of subgroups. In particular, the Texas study allowed for deeper examination of differences based on the language a survey was completed in (either English or Spanish) gender (as discussed earlier in the chapter), and respondents' management status. The initial findings suggest that future research should use workers' primary language as an independent variable for assessing working conditions and as a mediating or moderating variable in measuring other observed ethnic disparities. Future research should also examine mechanisms for career growth in the solar industry, as preliminary findings showed that White respondents were 42% likelier than Black respondents to manage other workers, 4% likelier than Hispanic respondents to manage others, and that those who submitted surveys in English were 21% likelier than those who submitted surveys in Spanish to carry out management responsibilities. This finding regarding management status is also complemented by the qualitative findings that indicate issues with favoritism and interplay between career opportunities and social networks among the interviewed workers.

## ACKNOWLEDGMENTS

Thank you to researchers at the Texas Climate Jobs Projects, Organized Power in Numbers, Social Science Research Solutions, and the Climate Jobs Institute. Thank you to the many workers, industry, government, and labor experts who gave crucial insights that led to the creation and design of these studies. Thank you to all the local organizations who helped distribute the surveys. Thank you to all the union organizers and union leaders diligently working to improve pay, benefits, and working conditions at solar jobsites. Last, thank you to the many workers who shared their on-the-ground perspectives with us. We appreciate your time and the opportunity to uplift your experience as we work together towards an equitable clean energy future.

## REFERENCES

Al Mubarak, Faisal, Reza Rezaee, and David A. Wood. 2024. "Economic, Societal, and Environmental Impacts of Available Energy Sources: A Review." *Eng* 5 (3): 1232–1265. https://doi.org/10.3390/eng5030067

Appel, Sam, and Jessie HF Hammerling. 2023. "California's Climate Investments and High Road Workforce Standards Gaps and Opportunities for Advancing Workforce Equity." https://tinyurl.com/437he45w

Behrani, Paras, Ahmad Shahrul Nizam Isha, Rohani Salleh, and Al-Baraa Abdulrahman Al Mekhlafi. 2023. "Determination of Occupational Health and Safety Risks in Solar Energy." *KnE Social Sciences* (December). https://doi.org/10.18502/kss.v8i20.14599

Boettner, Ted. 2024. "IRA Clean Energy Credits with Labor Standards Can Boost Union Jobs and Economy in Appalachia." https://tinyurl.com/dtj3rwa5

Buhai, I. Sebastian, and Marco J. Van Der Leij. 2023. "A Social Network Analysis of Occupational Segregation." *Journal of Economic Dynamics & Control* 147 (February): 104593. https://doi.org/10.1016/j.jedc.2022.104593

Combemale, Christophe, Dustin Ferrone, Patrick Funk, Amanda Quay, and Anna Waldman-Brown. 2024. "Workforce Analytic Approaches to Find Degrees of Freedom in the EV Transition." *SSRN Electronic Journal.* https://doi.org/10.2139/ssrn.4699308

Curtis, E. Mark, and Ioana Marinescu. 2023. "Green Energy Jobs in the United States: What Are They, and Where Are They?" *Environmental and Energy Policy and the Economy* 4 (January): 202–37. https://doi.org/10.1086/722677

Curtis, E. Mark, Layla O'Kane, and R. Jisung Park. 2024. "Workers and the Green-Energy Transition: Evidence from 300 Million Job Transitions." *Environmental and Energy Policy and the Economy* 5 (January): 127–161. https://doi.org/10.1086/727880

Davidson, Brendan. 2023. "Labour on the Leading Edge: A Critical Review of Labour Rights and Standards in Renewable Energy." *Energy Research & Social Science* 97 (March): 102928. https://doi.org/10.1016/j.erss.2022.102928

Denice, Patrick, and Jake Rosenfeld. 2018. "Unions and Nonunion Pay in the United States, 1977–2015." *Sociological Science* 5:541–561. https://doi.org/10.15195/v5.a23

Duroha, Jesse C., and Gretchen A. Macht. 2023. "Solar Installation Occupational Risks: A Systematic Review." *Safety Science* 160 (April): 106048. https://doi.org/10.1016/j.ssci.2022.106048

Dutta, N.S., H. Mirletz, B.K. Arkhurst, C. Houghteling, E. Gill, and S. Ovaitt. 2024. Applying Energy Justice Metrics To Photovoltaic Materials Research. *MRS Advances* 9, no. 12: 962–969.

Gadzanku, Sika, Alexandra Kramer, and Brittany Smith. 2023. "An Updated Review of the Solar PV Installation Workforce Literature." Technical Report. REL/TP—7A40-83652. https://doi.org/10.2172/1971876

Hoek Spaans, Avalon, and Jillian Morley. 2024. "Exploring the Conditions of the New York Solar Workforce." Cornell ILR Climate Jobs Institute. https://tinyurl.com/4wjw7p5j

Huckstep, Sam, and Helen Dempster. 2024. "Meeting Skill Needs for the Global Green Transition A Role for Labour Migration?" Center for Global Development. https://tinyurl.com/5n8d8z9b

Internal Revenue Service. 2025. "Frequently Asked Questions about the Prevailing Wage and Apprenticeship under the Inflation Reduction Act." https://tinyurl.com/yhk48wtz

Interstate Renewable Energy Council. 2024. "Solar Job Trends—Interstate Renewable Energy Council (IREC)." September. https://tinyurl.com/4c49jw5r

Lesser, Theodore. 2023. "Solar Strategies: Stakeholder Perspectives on a Chicago Solar Training Program's Intersectional Energy Justice Work." Knowledge@UChicago. https://doi.org/10.6082/UCHICAGO.6380

McAnaw Gallagher, Claire. 2022. "The Construction Industry: Characteristics of the Employed, 2003–20." Spotlight on Statistics. U.S. Bureau of Labor Statistics. https://tinyurl.com/fp2spbfx

Moe, Allison, Sika Gadzanku, Ryan Shepard, Heidi Tinnesand, and Elizabeth Gill. 2024. "Inventory of Clean Energy Education and Workforce Programs in Connecticut's I-91 Corridor." Technical Report. NREL/TP—5500-88569, 2368582. https://doi.org/10.2172/2368582

Mulvaney, Dustin. 2024. "Embodied Energy Injustice and the Political Ecology of Solar Power." *Energy Research & Social Science* 115 (September): 103607. https://doi.org/10.1016/j.erss.2024.103607

Parolin, Zachary, and Tom VanHeuvelen. 2023. "The Cumulative Advantage of a Unionized Career for Lifetime Earnings." *Industrial & Labor Relations Review* 76 (2): 434–460. https://doi.org/10.1177/00197939221129261

Solis, Miriam, Michael Oden, Katherine Lieberknecht, and Haijing Liu. 2025. "Labor Lacuna: Disjunctures Between Local Climate Action and Workforce Development in Advancing Just Transitions." *Journal of Urban Affairs* 21 (7): 2495–2517. https://doi.org/10.1080/07352166.2023.2291072

Srivastava, Rohini, Roxana Ayala, Robin Tuttle, and Allison Moe. 2024. "An Evaluation Framework for State Energy Offices' Energy Efficiency and Clean Energy Workforce Programs." Technical Report. NREL/TP—5500-88796, 2341269." https://doi.org/10.2172/2341269

Tabassum, Sanzana, Tanvin Rahman, Ashraf Ul Islam, Sumayya Rahman, Debopriya Roy Dipta, Shidhartho Roy, Naeem Mohammad, Nafiu Nawar, and Eklas Hossain. 2021. "Solar Energy in the United States: Development, Challenges and Future Prospects." *Energies* 14 (23): 8142. https://doi.org/10.3390/en14238142

Texas Climate Jobs Project, Cornell University School of Industrial and Labor Relations, Climate Jobs Institute, and Organized Power in Numbers. 2024. "Power and People Working Conditions in the Texas Clean Energy Transition." https://tinyurl.com/4z94werz

U.S. Bureau of Labor Statistics. 2023. "Geographic Profile of Employment and Unemployment, 2022." https://tinyurl.com/3r77rp3r

U.S. Bureau of Labor Statistics. 2024a. "47-2231 Solar Photovoltaic Installers." Occupational Employment and Wages, May 2023. https://tinyurl.com/2m9sr3hy

U.S. Bureau of Labor Statistics. 2024b. "Geographic Profile of Employment and Unemployment, 2023." https://tinyurl.com/446xtpk7

U.S. Bureau of Labor Statistics. 2024c. "Household Data Annual Averages." https://tinyurl.com/y3dw8dtt

VanHeuvelen, Tom, and David Brady. 2022. "Labor Unions and American Poverty." *Industrial & Labor Relations Review* 75 (4): 891–917. https://doi.org/10.1177/00197939211014855

Weil, David. 2005. "The Contemporary Industrial Relations System in Construction: Analysis, Observations and Speculations." *Labor History* 46, no. 4: 447–471. https://doi.org/10.1080/00236560500266258

# Organizing a Worker- and Community-Centered Transition: The Contra Costa Refinery Transition Partnership as Case Study

VIRGINIA PARKS
*University of California, Irvine*

JESSIE HF HAMMERLING
*University of California, Berkeley*

## Abstract

How do we ensure an equitable transition to a clean energy economy for workers and residents in fossil-fuel producing regions? We address the case of Contra Costa County, California, where the oil refinery sector provides an important source of well-paid, union jobs and local government tax revenue; it also has been a source of pollutants that have had serious health impacts for surrounding neighborhoods while contributing to greenhouse gas emissions. We focus on the work of a partnership of environmental justice and labor actors in Contra Costa County who have been working together to guide economic and climate policy and practice toward an equitable transition for workers and communities. We outline the research and organizing work guiding the initiative, the policy recommendations adopted by the partnership, and key outcomes of the initiative to-date. Key features include community- and worker-centered principles driving the process forward and the role of research in informing decision-making. We argue that key aspects of the initiative can inform other efforts, in other regions, seeking to achieve an equitable transition of workers and communities to a clean energy economy.

## INTRODUCTION

How do we build an equitable and sustainable economy as we decarbonize? This question motivates the growing consensus among activists, scholars, and policy makers that the hazards of climate change and economic inequality are inextricably linked. The United States has started to make progress toward understanding and acting on these interconnected challenges in the past decade through the Biden administration's Inflation Reduction Act and related policies, calls for a Green New Deal at the national level, and various state and local initiatives. Recent federal reversals of these initiatives aside, the path toward realizing a bold vision of transformational change that empowers workers and communities while reducing greenhouse gas emissions remains largely uncharted even in states with continuing decarbonatization programs. We contribute to the scholarly documentation of these emerging efforts with a case study of a partnership of labor and environmental actors in Northern California working to achieve an equitable clean energy transition. Our case highlights the regional economic development approach of this effort with a focus on the community- and worker-centered principles driving the process forward.

Our case focuses on the Contra Costa Refinery Transition Partnership (CCRTP), an initiative led by the BlueGreen Alliance that brings together labor and community organizations in the East Bay area of Northern California with the goal of generating and implementing a transition strategy that can meet the needs of workers, frontline communities, and the environment. Contra Costa County is a hub of fossil fuel production in California, home to four of the Bay Area's five oil refineries. When one of the refineries suddenly idled in 2020, laying off 750 workers, the disruptive effects of an unplanned transition became immediate and real for workers and residents. Although the closure of a refinery came as a relief to residents who have disproportionately suffered the ill health and environmental effects of the refinery industry for many decades, the prospect of the unknown effects of future changes in the refinery industry on the local region motivated actors to engage with one another around the question of transition planning.

Three features characterize the CCRTP initiative: (1) self-determination—the process is led by and engages actors most directly affected by the refinery industry, (2) collaborative deliberation informed by distributional justice—labor and community partners recognize that each other's harms require a differentiated response and redress, and (3) a regional economic development framework informed by shared priorities and values. We describe and discuss how research has played a vital informational role in advancing the work of the partnership. By providing actors with a shared basis of knowledge, they were better equipped to develop specific recommendations for achieving an equitable energy transition. Last, although our case reflects the local and variegated nature of the clean energy transition, we argue that key aspects of the initiative can inform efforts in other regions seeking to achieve an equitable transition of workers and communities to a clean energy economy.

## ORIGINS AND FORMATION OF THE CONTRA COSTA REFINERY TRANSITION PARTNERSHIP

The oil refinery sector in Contra Costa County provides an important source of well-paid union jobs and local tax revenue; it also has been a major source of pollutants that have had serious health impacts for surrounding neighborhoods and contributes to significant greenhouse gas emissions. California's climate policies—notably its phaseout of all sales of gasoline-powered vehicles by 2035—are necessarily changing the market for refinery products (California Air Resources Board 2021; California

Executive Department 2020). The industry has already shed employment, and further contraction is likely.

The first tangible impact of a changing refinery market was felt in 2020 when the Marathon Petroleum refinery in Martinez unexpectedly idled, throwing 750 people out of work. The impact on refinery workers and their families was devastating. The shuttered refinery was a warning to the region that similar closures were likely on the horizon. The union representing the Marathon Martinez refinery workers, United Steelworkers (USW) Local 5, realized that changes in the refinery industry could lead to future layoffs and an uncertain future for its members; it also realized that changes accompanying a clean energy transition would have widespread effects on the region beyond the fence lines of the refineries.

In the wake of the Marathon layoffs, the leadership of USW Local 5 reached out to local stakeholders to begin a conversation about what future decarbonization would mean for local workers and fence-line communities—residents living near industrial facilities. These discussions yielded the CCRTP initiative. The union engaged their longtime partners at the BlueGreen Alliance (a national labor–environmental alliance) to lead and facilitate the effort, along with the local labor federation (Contra Costa Labor Council), a local environmental justice organization (Asian Pacific Environmental Network, APEN), and the plumbers and pipefitters local union (UA Local 342) who perform significant maintenance work in the refineries. Additional support came from the University of California–Berkeley Labor Center and California Federation of Labor Unions. The goals of the CCRTP are to seek information about changes facing the refinery industry and their possible consequences for Contra Costa County, to build relationships and trust among key stakeholders, and to generate and implement a transition strategy that can meet the needs of workers, frontline communities, and the environment.

The successful creation of the partnership drew on long-standing relationships established through earlier campaigns addressing health and safety issues at the refineries that affect workers and residents stemming back to the 1990s. These groups had a precedent for collaboration. Organizing efforts in the aftermath of a Chevron refinery fire in 2012 were particularly influential in shaping the contemporary coalitional context and key to explaining the robustness of current transition initiatives in Contra Costa County.

In 2012, an explosion at the Chevron refinery in Contra Costa County led to a massive fire, putting workers' and residents' lives at risk (Goldberg 2018). Plumes of black smoke sent over 15,000 people downwind of the refinery to seek medical treatment after inhaling smoke and toxic fire gases. A coordinated organizing response ensued, bringing together labor and community organizations into the Refinery Action Coalition. APEN and the USW were key partners representing local fenceline community residents and refinery workers. Other groups included the BlueGreen Alliance, the Natural Resources Defense Council, the West County Toxics Coalition, and Communities for a Better Environment (Yoshitani and Ordower 2024).

These organizing efforts led to instrumental reforms to the refining industry. In particular, the BlueGreen Alliance played a lead role in coordinating efforts in close collaboration with the USW, which led to the nation's strongest refinery safety regulations, CalOSHA's Process Safety Management (PSM) regulations adopted in 2017 (BlueGreen Alliance 2017; California Department of Industrial Relations and California Environmental Protection Agency 2017). These regulations require refineries to invest in safety infrastructure and best-practices engineering and management. Workers are provided voice and an institutionalized frontline role in plant safety. The coalition's efforts ensured that labor and community actors were included in state development and deliberation of the PSM regulations; the state's Department of Industrial Relations convened public meetings attended by coalition members throughout the state and formed a labor–management technical committee to assist in drafting a proposal.

The ongoing organizing work that yielded these important safeguards to worker safety and public health set the stage for the formation of the current Contra Costa Refinery Transition Partnership. In the wake of the Marathon Martinez refinery closure, labor and community groups had a track record of organizing work to draw on. Bridge-building and co-deliberation are necessarily ongoing, but the groups' mutual respect and shared experiences facilitated the relatively quick collaborative commitment to an engagement process focused on the clean energy transition. Despite these groups' different views about the future of the refineries in the county and potential transition timelines, they have come together over shared concerns to plan for an equitable regional economy and sustainable local environment.

Through the efforts of the BlueGreen Alliance, the partnership received funding through the State of California's High Road Training Partnership (HRTP) program in 2022 to support the CCRTP's work to strategically analyze and develop an equitable transition agenda for the region (California Workforce Development Board, "High Road Training" and "Resilient Workforce Fund" no date). The HRTP program supports a diverse array of regional and industry partnerships aimed at addressing urgent questions related to income inequality, economic competitiveness, and climate change. The purpose of the state program is to foster collaborations that can deliver improvements in equity, sustainability, and job quality, while fostering economically and environmentally resilient communities across California (California Workforce Development Board, "High Road Training Partnerships," no date).

HRTP-funded CCRTP activities includes several research projects, along with a multi-year process of stakeholder engagement to inform and shape research products and to develop a comprehensive set of policy recommendations to facilitate a just transition in Contra Costa County.

## RESEARCH AND RECOMMENDATIONS FOR TRANSITION PLANNING

A key purpose of the CCRTP is to seek information about changes facing the refinery industry and their possible consequences for Contra Costa County. Partners expressed an urgent need to understand what changes lie ahead for the local refinery industry and what impacts these changes could have on the local environment and economy. First, the Marathon closure immediately reduced the number of well-paid, quality jobs for workers in the county without a college degree. Future loss of refinery jobs will further reduce these already limited opportunities. Second, refinery closures lead to a loss of tax revenue for local governments, threatening resources for essential services as well as public sector employment—also a source of quality, union jobs. Last, a poorly managed refinery transition could lead to ongoing public health and environmental hazards stemming from polluted sites, posing risks to workers, residents, and the environment. At the same time, refinery closures could present an opportunity for the region to improve public health outcomes and reduce inequality in the region, The CCRTP partners sought to define what just transition outcomes would look like and map out a strategy to achieve them.

The partnership commissioned a series of research reports to generate information addressing these issues. The research has had a twofold impact: in addition to providing stakeholders with invaluable information, the process of formulating research requests and reviewing results helped identify common ground and mutual concerns among partners. All stakeholders shared an initial starting point: they needed to know what changes in the refinery industry could mean for the region.

The BlueGreen Alliance served as the CCRTP's lead facilitator. The University of California Berkeley Labor Center served as the CCRTP's primary research partner. Center staff and affiliated researchers conducted original research to articulate challenges and opportunities related to refinery transitions in Contra Costa, then worked with key stakeholders to identify shared priorities for generating a framework to guide economic development planning toward an equitable transition. The BlueGreen Alliance facilitated a concurrent process with CCRTP partners and other key stakeholders in the region to develop a comprehensive set of policy recommendations, agreed on by all CCRTP members, that can ensure a just transition is achieved in Contra Costa.

The four reports released by the partnership are the following:

- "Fossil Fuel Layoff: The Economic and Employment Effects of a Refinery Closure on Workers in the Bay Area" (Parks and Baran 2023), which documents the layoff experience of Marathon oil refinery workers.
- "San Francisco Bay Area Refinery Transition Analysis" (Simeone and Lange 2025), which assesses scenarios for future changes in Bay Area refinery operations based on anticipated changes in demand for refinery products in California.
- "Refining Transition: A Just Transition Economic Development Framework for Contra Costa County, California" (Hammerling, Toaspern, and Schmahmann 2025), which synthesizes research findings on transition-related impacts on Contra Costa and proposes a stakeholder-informed approach for economic development.

- "Report and Recommendations of the Contra Costa Refinery Transition Partnership" (BlueGreen Alliance Foundation 2025), which summarizes the work of the partnership and presents a comprehensive set of CCRTP-approved recommendations to facilitate a just transition in Contra Costa.

We summarize the key findings of each report in turn, providing readers with an overview of the multiple stages and dimensions of the research effort in support of the CCRTP deliberative process. Each represents a stand-alone research effort carried out by independent researchers, yet the process of identifying research needs, shaping research requests, and reviewing results was an iterative process among partner members that served to determine and define shared interests and mutual goals. Learnings from one research project helped shape the next set of research questions and aims. The goal of self-determination within just transition processes was facilitated by a parallel process of inquiry and knowledge generation.

## Fossil Fuel Layoff: The Economic and Employment Effects of a Refinery Closure on Workers in the Bay Area

Marathon Petroleum's decision to idle its Contra Costa refinery in October 2020 took workers and their union, USW Local 5, by surprise. Beyond severance pay, which the union fought for in emergency bargaining sessions, the refinery closure thrust workers into an unplanned, unexpected transition with little warning and few resources—and with families to support.

The layoff was a wake-up call to the union, and it resolved to learn from the experience. As the first research project of the CCRTP, USW Local 5, the BlueGreen Alliance, and the UC Berkeley Labor Center identified the need for an academic study to assess the impact of the layoff on workers and their families and guide potential new policies and programs to minimize these impacts. Brainstorming about what could be learned, and how, from the Marathon experience sparked the idea for a worker survey. UC Berkeley Labor Center staff reached out to Virginia Parks to design and lead the research project, funded through the CCRTP's HRTP grant from the State of California.

The speed of the survey project belies an intensive process of discussion and deliberation between the primary investigator (PI) and the union to achieve parallel priorities: a scientifically rigorous and robust research process for the PI and a transparent, democratic decision-making process for union leadership. After a series of conversations between the PI and the union president about survey methods, research ethics, and the Institutional Review Board process (e.g., the requirement of "noninterference" on the part of the union), the union voted on whether to go forward with the research project—first with its executive board and then with its full membership. The research moved forward only after receiving approval through these democratic union decision-making processes, all on the heels of an abrupt layoff.

Each of the 345 laid-off permanent refinery workers were contacted with information about the survey in December 2021. The survey remained open throughout the first week of March 2022. The survey response rate was 41% (n = 140). Twenty-one workers participated in follow-up Zoom or telephone interviews (see Parks and Baran 2024 for more details about methods).

The survey confirmed the local reputation of refinery jobs as well-paid, union positions with safe working conditions. Decades of union representation yielded high wages, a vigilant safety culture, and a strong occupational identity among refinery workers. Job benefits included robust retirement plans, family health benefits, and several weeks of paid vacation. The union facilitated a worker-to-worker model of formal and informal training across a broad scope of refinery operations, including basic protocol, health and safety, and emergency response. Workers could start at the refinery with few skills but develop a highly specialized skill set over time. Workers were guaranteed pay increases for job tenure and job advancement. The union provided additional training and leadership opportunities, as well as a democratic vehicle for decision making and input. As a result, workers made lifetime careers at the refinery.

Following the refinery closure in October 2020, laid-off refinery workers were relatively successful in securing post-layoff employment but at a cost. Median hourly wages were 24 percent lower than the median refinery wage. The median hourly wage at the refinery was $50, compared to a post-layoff median of $38. More than a quarter of all workers remained unemployed a year after layoff. Workers also reported more difficult working conditions at post-layoff jobs. They described hazardous worksites, heavy workloads, work speedup, increased job responsibilities, and few opportunities for advancement. Above all, workers cited poor safety practices and increased worksite hazards as the most significant and alarming characteristics of degraded working conditions.

The process of finding these new jobs posed unforeseen challenges for workers. No coordinated job search assistance was available to workers through public or nonprofit programs. The State of California's workforce assistance consisted of an online inventory of job advertisements that lacked a nuanced search function, yielding "forklift operator" positions when laid-off refinery operators searched for jobs matching their former occupations. Workers described two unexpected challenges: (1) employers' lack of knowledge about refinery work and refinery workers' skills and (2) the inability of workers to prove their skill or experience through certifications or a verification. Most of these blue-collar workers did not have a college or specialized educational degree to show employers, nor do certification programs exist tied to their refinery positions. As a result, workers struggled to verify their skills and competencies to future employers. Last, workers reported anxiety about future economic stability. A full third of all workers described that they were "falling behind financially" a year following the layoff compared to only 3% before the layoff. Workers reported struggling with depression and anxiety.

The survey confirmed what large-scale displaced worker surveys have repeatedly found: workers experienced post-layoff wage decreases, and many experienced longer-term periods of unemployment. But the survey also revealed surprises. Workers experienced unexpected friction in the job-skills matching process due to the inability of potential employers to discern job candidates' skills. Workers' lack of formal skill credentials combined with the employers' lack of knowledge about refinery jobs created an information vacuum that puts workers at a disadvantage in the labor market. The survey findings lend insights and warnings for policy makers engaged in transition planning. Two bear emphasis: (1) workers want re-employment in well-paid, quality jobs, regardless of industry (e.g., a one-to-one match into the renewable energy sector should not define transition planning) and (2) workers are interested in job training that leads to good jobs in an efficient and timely manner.

The survey research represents an information-gathering and knowledge-building process that centers workers and their experiences. It reflects the CCRTP's guiding principle that transition planning must engage and be informed by the experiences of those most directly impacted by refineries and the clean energy transition. Notably, the survey findings directly influenced the funding parameters of the State of California's Displaced Oil and Gas Worker Fund (DOGWF). The state's Employment Development Department incorporated recommendations from the Marathon worker survey report when crafting its funding parameters and request for proposals for DOGWF funds, a pilot fund intended to support oil and gas workers displaced by the state's efforts to transition to renewable energy (California Legislative Information 2022). Funding requirements were released in the fall of 2023 that included training and certification initiatives as well as general support and counseling services. In February 2024, $26.7 million was awarded from this fund to workforce development agencies, training organizations, and labor unions, including $9.8 million to the Steelworkers Charitable and Education Organization, the nonprofit entity of the United Steelworkers, to create a statewide program to support impacted oil workers (California Employment Development Department 2024).

## San Francisco Bay Area Refinery Transition Analysis

CCRTP partners wondered whether the layoffs at Marathon were a one-off or a harbinger of imminent closures at Bay Area refineries. All parties lacked a reliable source of information about what changes lay ahead, so the CCRTP commissioned an analysis of refinery transition scenarios to better understand the changing market for refined petroleum products in California and how the local refineries would likely respond. The "San Francisco Bay Area Refinery Transition Analysis," led by energy economists Christina Simeone and Ian Lange, was completed in 2022 and released concurrently with the CCRTP's summary report and recommendations in 2025.

California's ambitious climate policies are well known; the state has led the nation in its policies to limit greenhouse gas emissions and reduce consumption of fossil fuels, beginning with the passage of the landmark 2006 legislation AB 32, The California Global Warming Solutions Act. In 2020, Governor Newsom signed Executive Order N-79-20, which requires that all new cars sold in the state by 2035 must be zero-emission vehicles, later codified with a phased-in approach by the California Air Resources Board's (CARB) Advanced Clean Car II rule. There is no doubt that these policies are having and will continue to have consequences for the state's oil refineries (which primarily produce gasoline and diesel fuels), but questions remained about the likely outcomes at Contra Costa refineries.

The "Refinery Transition Analysis" found that the Bay Area is likely to see major changes in refinery operations by 2045, including significant reductions in output along with facility closures. Simeone and Lange estimated future demand for Bay Area refinery products using the transportation energy demand projections included in the 2022 CARB Final Scoping Plan. The analysis found that the reduced market demand for gasoline by 2045 will lead to a reduction between 65% and 92% in the required refinery capacity in the Bay Area, with variation depending on demand for exports. While the precise timing and process of refinery transition in the Bay Area is impossible to predict with certainty, overall expected decline in demand for refinery products will threaten the viability of traditional petroleum refinery operations in the Bay Area and is likely to lead to facility closures.

Sharing the findings of the "Refinery Transition Analysis" with CCRTP partners established a common understanding and acceptance of the overall trajectory of refinery transition in Contra Costa. Partners continue to maintain differing perspectives about the policies that affect how and how fast changes in the refinery market are likely to unfold, but the findings of the "Refinery Transition Analysis" made it clear that market changes resulting in further contraction in the industry have already been set into motion.

## Refining Transition: A Just Transition Economic Development Framework for Contra Costa County, California

The petroleum industry is a major contributor to economic activity in Contra Costa County. In 2022, petroleum refining in the county produced over $42 billion in output, representing 24% of Contra Costa County's total output. The likelihood of additional refinery closures in Contra Costa necessitates an understanding of the refinery industry's current role in the regional economy, to understand the impacts of closures and to inform planning processes.

The University of California–Berkeley Labor Center completed a comprehensive analysis of the economic impact of the refinery industry on Contra Costa County, including employment projections and a tax impact analysis. We summarize key findings of their "Refining Transition" report (Hammerling, Toaspern, and Schmahmann 2024) here and direct readers to the report for a description of the methodologies employed.

The report also includes a proposed just transition framework that can be used by nongovernmental stakeholders and policy makers to inform economic development planning to respond to the specific conditions of transition in Contra Costa County. Its purpose is to provide the architecture for a community- and worker-led planning process that can help move the local economy away from low-road economic activities toward a sustainable, high-road economy characterized by good jobs, community health, and economic resilience.

*Employment*

The report contextualizes transition against the trend of rising income inequality experienced by Contra Costa County residents over the past several decades. Inequality has been exacerbated by the county's decline of middle-wage, unionized jobs and the growth of low-wage, nonunion jobs. Reflecting national trends, union membership has plummeted in the region: in 2023, 8% of private sector workers in the Bay Area were represented by a union, compared to 18% in 1986.

Paralleling this trend is the rise of service sector jobs and the decline of industrial jobs. Manufacturing jobs account for 4.5% of all jobs in Contra Costa County, or 15,000. Petroleum and coal products manufacturing, a subsector of manufacturing, accounts for just under 1% of total employment. The two largest industries of employment in Contra Costa County are healthcare and social assistance (18.5% of total employment, with 60,000 jobs) and retail trade (13.7% of total employment, with 45,000 jobs). Occupations with the greatest projected growth are lower skilled and lower paid, such as home health and personal care aides, fast-food and counter workers, cooks (restaurant), and laborers and freight, stock, and material movers. Currently, the top occupations by total employment in Contra Costa County that do not require a college degree earn a median wage around $20 per hour (home health and personal care aides, cashiers, and fast-food workers).

By contrast, the refinery industry has been an important source of well-paid jobs in Contra Costa County for workers without a college degree. Through the ongoing organizing efforts of workers, the refineries provide a source of well-paid, union jobs. Workers collectively bargain over wages, benefits, and working conditions. As a result, jobs in the refinery industry—along with the unionized construction trades—are among the highest-paying jobs in Contra Costa County that do not require a four-year degree. Petroleum pump system operators, refinery operators, and gaugers earn a median wage of $50 per hour, not including benefits.

Yet the refinery industry in Contra Costa County will clearly contract, even if the timeline is uncertain. Transition in the refining industry is already under way, as the Marathon refinery idling illustrated. Between 2012 and 2021, the industry experienced a 21% decline in employment. Current growth forecasts are sobering. The report's job-match analysis shows that refinery workers have skills that could translate to a wide range of occupations and industries, including other types of manufacturing, facilities operations and maintenance, and construction. However, job growth in these industries is currently projected to be limited, and wages in comparable occupations are around 50% lower than in refinery occupations. Overall, the greatest growth in the county is predicted for the lowest-wage occupations. The report emphasizes that improving the quality of these jobs will be particularly important for Contra Costa County.

*Taxes*

A core component of the report is an empirical analysis of the tax impacts of the refining industry on the county. Importantly, this analysis provided a shared lens for multiple and diverse stakeholders to view and evaluate the scope of effects that will accompany transition. Taxes paid to local governments help fund essential public services, including infrastructure, public safety, sanitation, water, and social services. Good-quality public sector employment also depends on these tax revenues. As petroleum refining declines in Contra Costa, tax revenue generated by the industry and its supply chain will also decline.

Combined tax revenue from the refinery industry and its supply chain currently constitutes a substantial share of local government revenue in Contra Costa County. The "Refining Transition" report identifies tax revenue generated by the refinery industry in Contra Costa County via sales taxes, property taxes, special assessments, and other taxes paid by refinery companies (direct tax impacts). Goods and services traded with the refinery industry along its supply chain also generate tax revenue (indirect tax impacts). The analysis found that the refinery industry in Contra Costa directly contributed an estimated $136 million in local taxes in 2022; activities that provide inputs into the refinery industry in Contra Costa contributed an estimated $836 million in local taxes in 2022. Combined direct and indirect local tax impacts in Contra Costa generated an estimated cumulative local tax revenue of $972 million.

The report found that the refinery supply chain in Contra Costa County contributes a much larger share of local tax revenue than the refinery companies (86% indirect vs. 14% direct impacts). Local taxes paid directly by the refinery companies are relatively low compared to their share of total economic output. The report states,

> Community advocates have argued that the refinery industry ought to pay higher taxes than they currently do, based on the substantial environmental and health externalities associated with the operation of the facilities, the property tax limits of California's Proposition 13, and the increasing profits of oil refinery companies. (Hammerling, Toaspern, and Schmahmann 2024: 37)

Even at current levels, the cumulative taxes paid by the refinery industry and its supply chain are substantial. As the industry shrinks, so too will this tax base.

### Regional Economic Development Planning Framework

The second component of the "Refining Transition" report lays out a just transition economic development framework developed in consultation with the partners to guide transition planning in Contra Costa County. This framework emerged from a stakeholder-guided process and provides an instructive template for just transition efforts in other regions.

A definitional aspect of the CCRTP is its deliberative process of setting priorities and guidelines to inform economic development planning. This process involved regular and ongoing meetings among partners governed by principles of co-determination and collaborative engagement. The report acknowledges this process:

> The partners involved in the CCRTP represent the constituencies most affected by refinery transition and its consequences, in

some similar ways and in some very different ways. Their shared priorities represent important areas of agreement, where a broad array of stakeholders can collaborate and problem-solve around goals held in common for the county. (Hammerling, Toaspern, and Schmahmann 2024: 46)

Notably, research played an instrumental role in focusing and informing these discussions.

Part 2 of the report represents the outcomes of these discussions as well as the process through which these outcomes were produced. Informed by the findings of the first part of the report's research, CCRTP partners identified three primary challenges posed by transition in the refinery industry as starting points for economic development planning efforts:

- Loss of tax revenue for essential services
- Loss of good jobs without comparable replacements
- Continued environmental and public health hazards from disinvestment, climate change, and new industries (Hammerling, Toaspern, and Schmahmann 2024: 41–42)

These starting points led to the CCRTP's adoption of three just transition priorities for Contra Costa County:

- Good jobs: Increase quality jobs and ensure those jobs are accessible to those who need them most, including displaced refinery workers and workers from historically marginalized communities.
- Healthy communities: Stop polluting land, air, and water in communities that have had disproportionate pollution burdens and target environmental and health improvements in these areas.
- Economic resilience: Reduce economic dependency on fossil fuel production by diversifying and increasing other sources of tax revenue.

Economic development activities that advance all three of these objectives should be prioritized in economic development planning and promotion efforts. These priorities are followed by a set of just transition principles: (1) commit to democracy by giving worker- and community-based organizations a central role in transition planning and supporting workplace democracy by supporting unions in the local economy, and (2) pursue demand-driven workforce strategies that generate just transition outcomes, particularly in shaping job quality and access.

The remaining sections of the "Refining Transition" report identify a set of tools and investment strategies that local governments could adopt to achieve these three objectives by encouraging and rewarding high-road growth while closing off low-road economic activities. Examples include negotiated agreements such as project labor agreements, community benefits agreements, and card-check neutrality agreements; standards linked to public investments such as prevailing and living wage requirements; and responsible contractor criteria and apprenticeship and skills standards and certifications. We encourage readers to consult the report for more examples of specific tools and additional resources.

## Report and Recommendations of the Contra Costa Refinery Transition Partnership

Over the course of the three-year project, BlueGreen Alliance convened the CCRTP for monthly meetings to discuss research process and findings, and to develop a shared set of policy recommendations. The final "Report and Recommendations of the Contra Costa Refinery Transition Partnership" summarizes the work and research of the CCRTP and presents the 31 recommendations endorsed by the partners.

The process of developing the recommendations involved structured input from local stakeholders outside of the CCRTP. The CCRTP convened two working groups, a Worker Safety Net and Transition Working Group and a Community Solutions Working Group; organized a three-session workshop series with frontline community members in Richmond, California; and conducted multiple sets of briefings and feedback sessions with a wide range of additional community, labor, and government stakeholders.

The recommendations are grouped into five categories, all aimed at state and local policy makers:

- Refinery transition planning and oversight
- Refinery worker safety net and transition
- Land use, decommissioning, and cleanup
- Refinery community support and transition
- Just transition economic development

Examples include requiring refineries to provide two-year advance notification of closure or conversion, creating a refinery transition oversight commission, establishing refinery decommissioning and cleanup standards, establishing local community recovery and transition funds, establishing an energy tax remediation fund, and integrating community-centered principles into economic development planning processes. We encourage readers to consult the full set of recommendations as examples of how organizations most directly impacted by the energy transition have moved from general calls for a just transition to specifying exactly *how* such a transition can be achieved through policy formulation.

## CONCLUSION

On the organizing and political mobilization front, the CCRTP has created an independent, organized voice for workers and community members most directly affected by transition in the refinery industry—a critical goal of a just transition. The CCRTP coordinates with government agencies and decision makers and seeks to shape local policy agendas and policy outcomes. Yet the partnership remains independent in ways critical to maintaining its commitment to self-determination. In this chapter, we highlight how research facilitates and moves forward this principle. Through identifying research needs, shaping research requests, and reviewing results, partnership members collaboratively determined and defined shared interests and mutual goals. By defining research questions, the partnership initiated research inquiry that centered the experiences of workers and residents most directly affected by transition. The approach of the CCRTP illustrates how the goal of self-determination within just transition processes is moved forward by a parallel process of inquiry and knowledge generation.

A second defining feature of the CCRTP is a commitment to collaborative deliberation informed by distributional justice. This is a commitment that requires constant vigilance and time. In the case of the CCRTP, time comes in many forms. First, labor and community organizations had a history of campaign organizing to draw on—shared efforts to transform the refinery industry in response to past public health and worker safety violations. Second, partnership members meet regularly and often. Decisions are made at the speed of trust—whatever time is needed to build trust is taken. Last, partners recognize that each organization represents members with different experiences of harm and privilege. Redress and benefits of a just transition will necessarily be differentiated. Despite differing views about the fate of the refinery industry in the region and the specific timeline of transition, these labor and community organizations continue to work together to identify shared goals.

A third defining feature of the CCRTP is its adoption of a regional economic development framework through which to analyze transition and develop policy recommendations. This framework facilitates both a comprehensive and locally focused transition planning process. Again, research has played a critical role on this front by providing partners with a shared basis of information from which they can draw on to develop both specific and transformative recommendations for achieving an equitable energy transition. At the time of writing, the "Refining Transition" report had just been released and its influence on local economic development planning initiatives not yet known. However, the ambitious and specific policy recommendations released by the CCRTP in early 2025 reflect an intensive process of debate and, ultimately, agreement among labor and community partners for how a just transition in the Contra Costa region can be achieved—a consequential and promising step toward building an equitable and sustainable economy as we decarbonize.

## REFERENCES

BlueGreen Alliance. 2017. "After Five-Year Effort, California Adopts the Nation's Strongest Refinery Safety Regulations." May 18. https://tinyurl.com/bddb8h77

BlueGreen Alliance Foundation. 2025. "California Contra Costa Refinery Transition Partnership Report and Policy Recommendations." January 14. https://tinyurl.com/22dmmp2a

California Air Resources Board. 2021. "Governor Newsom's Zero-Emission by 2035 Executive Order N-79-20." January 19. https://tinyurl.com/nfzj5anh

California Department of Industrial Relations and California Environmental Protection Agency. 2017. "New Regulations Improve Safety at Oil Refineries." August 4. https://tinyurl.com/vnd7urdp

California Employment Development Department. 2024. "The EDD Awards $26.7 Million to Provide Job Training and Support Services to Displaced Oil and Gas Industry Workers." February 8. https://tinyurl.com/22p5evaz

California Executive Department. 2020. "Executive Order N-79-20." https://tinyurl.com/wdja7suu

California Legislative Information. 2022. "Unemployment Insurance Code §9920-9925. Article 7. Displaced Oil and Gas Worker Pilot Program." https://tinyurl.com/2cex3vyw

California Workforce Development Board. No date. "High Road Training Partnerships." https://tinyurl.com/2us8545x

California Workforce Development Board. No date. "High Road Training Partnership: Resilient Workforce Fund Award Announcement." https://tinyurl.com/5n767udj

Goldberg, Ted. 2018. "Chevron, Richmond Settle Lawsuit Over 2012 Refinery Fire." KQED. May 3. https://tinyurl.com/3m2drt75

Hammerling, Jessie HF, Will Toaspern, and Laura Schmahmann. 2025. "Refining Transition: A Just Transition Economic Development Framework for Contra Costa County, California." January 14. UC Berkeley Labor Center. https://tinyurl.com/2sewtfbr

Parks, Virginia, and Ian Baran. 2023. "Fossil Fuel Layoff: The Economic and Employment Effects of a Refinery Closure on Workers in the Bay Area." April 26. UC Berkeley Labor Center. https://tinyurl.com/mr8dxkwz

Parks, Virginia, and Ian Baran. 2024. "Fossil Fuel Layoff: Capturing Decarbonization Impacts on Workers Through a Refinery Closure." *Labor Studies Journal* 50, no. 1: 28–53. https://doi.org/10.1177/0160449X241281938

Simeone, Christina, and Ian Lange. 2025. "San Francisco Bay Area Refinery Transition Analysis." January 14. The BlueGreen Alliance Foundation. https://tinyurl.com/22dmmp2a

Yoshitani, Miya, and Jeff Ordower. 2024. "Introduction: Our Future Story." In *Power Lines: Building a Labor–Climate Justice Movement*, edited by Jeff Ordower and Lindsay Zafir. The New Press.

# CHAPTER 4

# Fracking or No Fracking?
# How a Green Transition Can Work for Workers

ROBERT POLLIN
JEANNETTE WICKS-LIM
*University of Massachusetts Amherst*

## ABSTRACT

Throughout the 2024 U.S. presidential campaign, the only climate-related issue to achieve prominence was the question of whether to ban fracking operations in the United States. Donald Trump is a long-time climate denier and therefore had no qualms in supporting fracking and all other techniques for extracting fossil fuels from the ground. By contrast, Kamala Harris had supported a nationwide ban on fracking during her 2019 presidential campaign. This was because of the severe negative environmental and public health impacts of this natural gas extraction technique and because burning natural gas to produce energy generates $CO_2$ emissions that cause climate change. But Harris opposed a fracking ban in 2024 on the assumption that the ban would impose major costs to the economy of Pennsylvania, which has the second-largest fracking operations among U.S. states, after Texas only. Such negative economic outcomes in Pennsylvania would indeed result if fracking were banned in the United States and no large-scale alternative economic activities were introduced into Pennsylvania's economy. But banning fracking must be understood as one component of a much larger program to advance a viable climate stabilization program in Pennsylvania and everywhere else. We find that building a clean energy–dominant infrastructure in Pennsylvania—focused on investments in energy efficiency and renewable energy sources—will generate approximately 160,000 jobs in the state. Meanwhile, phasing down fracking and all other fossil fuel activities by 50% between 2026 and 2035 will entail job losses in the range of 1,700 per year within the state. We argue that these 1,700 displaced fossil fuel workers should receive just transition policies that include pension, employment, and income guarantees, in addition to, as needed, retraining and relocation support. We estimate that such a just transition program for these workers will cost in the range of $240 million per year. This amounts to about 0.02% of Pennsylvania's current GDP. Thus, we show how, between 2026 and 2035, Pennsylvania could phase out 50% of all its fossil-fuel production activities—including fracking operations—while also providing generous support for workers to transition out of their fossil fuel–industry jobs and into activities that both raise public health and environmental standards in the state and contribute toward a viable global climate stabilization project.

## INTRODUCTION

The deepening climate crisis was almost entirely ignored as an issue throughout the 2024 U.S. presidential campaign between Donald Trump and Kamala Harris. In fact, only one climate-related issue achieved prominence during the campaign. This was the question of whether to ban fracking operations in the United States (Lefebre 2024).

"Fracking" is an informal term of reference for hydraulic fracturing. This is a technology used to extract oil and natural gas from underground rock formations, such as sandstone, limestone, or shale rock deposits. Fracking technology is employed as a means of increasing the rate at which oil and gas can be extracted profitably from such rock formations. But fracking operations also generate severe environmental and health impacts through water and soil contamination as well as noise pollution. This is why, as of 2024, five U.S. states had banned fracking, including California, New York, Washington, Maryland, and Vermont.

Donald Trump has long been a rapturous cheerleader on behalf of fossil fuel production in the United States, regularly invoking the "drill baby drill" slogan first popularized by the 2008 Republican vice-presidential candidate Sarah Palin. It follows that Trump would have no qualms whatsoever in supporting fracking and all other techniques for extracting fossil fuels from the ground. This is the case, even though burning oil, coal, and natural gas to produce energy is, by far, the most significant driver of the climate crisis, because burning fossil fuels releases carbon dioxide ($CO_2$) emissions into the atmosphere. The accumulated stock of $CO_2$ emissions in the atmosphere, in turn, is the most important factor that produces climate change. Trump, of course, denies these most basic conclusions established by climate science. Since taking office in January 2025, President Trump himself and his administration more generally have reiterated their unqualified support for expanding fossil fuel production within the United States—that is, "drill, baby, drill"—and their vehement opposition to "green new deal social engineering programs."[1]

Kamala Harris, by contrast, has always made it clear that she supports the findings of climate science. She also supported a nationwide ban on fracking during her first presidential campaign in 2019. Harris then dropped her opposition to fracking when she became the vice-presidential candidate in Joe Biden's 2020 presidential campaign and maintained that position throughout the Biden–Harris administration of 2021–2024. Throughout her 2024 presidential campaign, Harris continued to support fracking operations in the United States. This was true even though Trump regularly reminded voters of Harris's earlier opposition to fracking and claimed that Harris would end up banning fracking if she were elected president.

Harris did have an obvious strategic, if unprincipled, motivation for maintaining her support for fracking during the 2024 election campaign. This was because of the centrality of Pennsylvania in determining the election's outcome—that is, Pennsylvania was, at once, the most important contested state in the 2024 campaign, as well as the state with the second-largest fracking operations in the United States, following only Texas. Thus, supporting a ban on fracking was understood to be a strategic nonstarter for Harris's campaign. The assumption was that a ban on fracking would inflict major damage on Pennsylvania's economy.

Such negative economic outcomes in Pennsylvania would indeed result if fracking were banned in the United States and no large-scale alternative economic activities were introduced into Pennsylvania's economy. But, in fact, banning fracking cannot be understood as an isolated, one-off policy measure. Rather, it should be recognized as one component of a much larger program to advance a viable climate stabilization program in Pennsylvania and everywhere else.

Focusing on Pennsylvania, a fracking ban would be one component of an overall program to achieve zero $CO_2$ emissions in the state, by phasing out fossil fuel consumption and building a high-efficiency, renewable energy–dominant infrastructure as the alternative to the state's existing fossil fuel–dominant infrastructure.[2] The research we review in this chapter shows that the investments needed to build this alternative clean energy infrastructure in Pennsylvania will generate far more jobs in the state than the jobs that will be lost through phasing out fracking and all other fossil fuel–based activities. Moreover, we will also show that the most viable clean energy transition program for Pennsylvania is one that incorporates just transition policies for the workers and communities in the state that are currently dependent on the fossil fuel industry for their livelihoods. This overall package of measures could indeed be characterized as major components of a "Green New Deal" program for Pennsylvania in particular, as well as for the United States and global economies more generally.[3]

Specifically, we will discuss here how building a clean energy–dominant infrastructure in Pennsylvania—focused on investments in energy efficiency and renewable energy sources—will generate approximately 160,000 jobs in the state. Meanwhile, phasing out fracking and all other fossil fuel activities will entail job losses in the range of 1,700 per year within the state over a 10-year period, 2026–2035, in which fossil fuel consumption would contract by 50%. This decline in fossil fuel consumption would be matched by a comparable decline in fossil fuel production activity in the state.

Policies will also certainly need to be enacted to provide robust transition support for these 1,700 displaced workers per year—what the late U.S. labor leader and environmentalist Tony Mazzocchi termed "just transition" policies. As early as 1993, Mazzochi wrote:

> Paying people to make the transition from one kind of economy—
> from one kind of job—to another is not welfare. Those who work
> with toxic materials on a daily basis … in order to provide the world

with the energy and the materials it needs deserve a helping hand to make a new start in life.

The critical point in Mazzocchi's idea is that providing high-quality adjustment assistance to today's fossil fuel–industry workers will represent a major contribution toward making a global climate stabilization project viable. Without such adjustment assistance programs operating at a major scale, the workers and communities facing retrenchment will, predictably and understandably, fight to defend their communities and livelihoods.

The just transition policies that we propose for Pennsylvania's fossil fuel industry–based workers include job guarantees, wage insurance, and pension guarantees, as well as job placement, training, and relocation support. We estimate that the costs of providing this level of just transition support will cost about $240 million per year. This would amount to roughly 0.02% of Pennsylvania's average GDP between 2026 and 2035.

As such, when included as one component of an overall clean energy and climate stabilization program, banning fracking in Pennsylvania will achieve significant environmental and health benefits for the state's residents while contributing toward eliminating $CO_2$ emissions—that is, eliminating the single largest factor causing the global climate crisis. This overall clean energy and climate stabilization program will also become a major engine for expanding job opportunities and increasing living standards throughout the state, generating far greater net benefits than what has been achieved through the state's fracking operations.

The rest of this paper proceeds as follows. In the next section, we review the economic, public health, and environmental impacts of fracking operations in Pennsylvania. After that, we present an outline of an emissions reduction and clean energy expansion program for the state. We estimate that about 160,000 jobs will be generated within Pennsylvania through a clean energy investment program scaled at about $26 billion per year, equal to about 2.5% of the state's GDP. Then we review the clean energy investments that have resulted in Pennsylvania in connection with the two major Biden-era programs, the Inflation Reduction Act (IRA) and the Bipartisan Infrastructure Law (BIL). Investments in Pennsylvania associated with these two programs totaled about $9 billion at the end of 2024.

In the last section, we focus on the job losses that will result in Pennsylvania through the fossil fuel phase-out in the state and just transition policies to support the workers who will experience displacement through this phase-out. We estimate that about 1,700 fossil fuel–based workers per year will experience displacement through a ten-year program to reduce fossil fuel production and consumption in Pennsylvania by 50%. This total figure for displaced workers includes workers engaged in fracking operations, as well as all other fossil fuel–based activities. We propose that support for all these workers should include pension, employment, and income guarantees, as well as retraining and relocation assistance as needed. The chapter concludes with some general observations on how a green transition program in Pennsylvania and elsewhere can support working people—including those now employed in fracking operations and other fossil fuel–

based activities—while also driving down $CO_2$ emissions and thereby advancing a viable climate stabilization project.

## IMPACTS OF PENNSYLVANIA'S FRACKING OPERATIONS

Pennsylvania is a major supplier of fossil fuel energy in the United States. As of the most recent 2022 data, it ranks second among U.S. states, behind only Texas, in producing natural gas, providing 17.5% of overall U.S. gas supply. It also ranks third, behind Wyoming and West Virginia, in producing coal, contributing 8.7% of overall U.S. production.[4]

The most significant fossil fuel energy source in Pennsylvania is the Marcellus shale formation. In 2008, natural gas began being extracted from Pennsylvania's Marcellus Shale deposits on a large scale through fracking operations along with horizontal drilling technology. The Marcellus formation extends under three-fifths of Pennsylvania as well as parts of West Virginia, New York, Ohio, and Maryland. But most of the gas extraction activity in the Marcellus shale formation is concentrated in northeastern and southwestern Pennsylvania. This is both because the gas deposits in these parts of Pennsylvania are relatively accessible through fracking technology and because the policy framework in Pennsylvania has supported fracking. In contrast, as noted above, New York and Maryland have banned fracking operations to date, even though those states have potentially significant gas reserves to exploit in the portions of the Marcellus shale that are within their respective borders.

We review here the experience with fracking operations in Pennsylvania since their inception in 2008, in terms of their economic, environmental, and public health impacts.

### Economic Impacts

Between 2007 and 2014, employment grew strongly in northeastern and, to a somewhat lesser extent, southwestern Pennsylvania as a result of the fracking boom. But the employment level peaked in 2014 and has experienced a declining trend subsequently. In Figure 1, we show statewide employment data for three industries in the state—oil and gas extraction, drilling oil and gas wells, and support activities for oil and gas operations. As we see, employment in these oil and gas industries increased fourfold between 2007 and 2014, from 5,829 to 23,525. The employment level then began falling off after 2014. As of the most recent 2023 data, statewide employment in these oil and gas sectors was at 12,156, a nearly 50% decline relative to the 2014 peak.

Average real wages in oil- and gas-related employment also rose sharply in the initial phase of the state's fracking boom, from the 2007 level of about $90,000 (in 2023 dollars) to about $127,000 in 2017, a 41% increase. But wages then began a general, if uneven, downward trajectory. As of 2023, average real wages were at $112,691, an 11% decline relative to the 2017 peak.

Since 2019, the industry has experienced a sharp slump, as reflected especially in the data on employment decline. This decline began prior to the 2020 COVID-

Figure 1. Employment Level and Real Wages for
Pennsylvania Oil and Gas Operations, 2007–2023

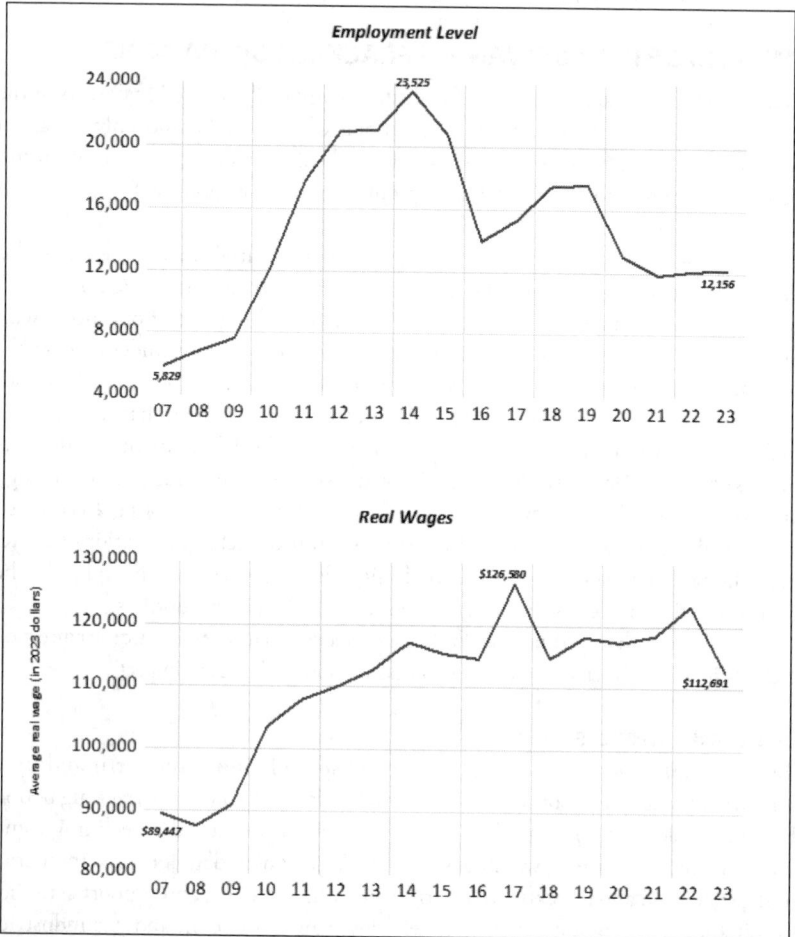

U.S. Bureau of Labor Statistics, QCEW
Note: Figures are for oil/gas extraction, drilling and support activities.

induced recession. But the recession deepened the contraction. Fracking operations in the state have not since returned to their 2014 levels.

## Public Health and Environmental Impacts

A large number of reports of the negative environmental and public health impacts of fracking operations in Pennsylvania began to emerge soon after these operations expanded to a large scale in the state in 2008. Thus, in June 2020, an Allegheny County grand jury report documented these impacts in detail, based on two years of research and direct testimony.[5]

A summary of some of the findings of the grand jury report includes the following passages:

> Wells can be drilled as close as 500 feet from your front door. Once construction of a well pad begins, life changes. We heard about the clouds of dust, the grimy film, the booming and the blinding lights, day and night. The construction phase of the process is still just the beginning. Next comes the drilling and the hydraulic fracturing of the wells. These parts of the process bring their own nuisances, some of which are similar to what homeowners experienced during the construction phase. Oftentimes, the noise is far worse than it was during the construction phase and can occur 24 hours a day. Some people had to sleep in a corner of the basement trying to get away from it. …
>
> Aside from the nuisances of the process, some people, as we learned from testimony, began to notice changes to their water. In many areas where unconventional oil and gas activity is common, there is no public water line. People rely entirely on water wells drilled on their own property. When the oil and gas operators spilled products used to fracture a well, or the storage facilities that held the wastewater leaked, the chemicals made their way into the aquifers that fed those water wells. The water started smelling like sulfur, or tasting like formaldehyde. It burned the skin. …
>
> Then there was the air. The smell from putrefying waste water in open pits was nauseating. Airborne chemicals burned the throat and irritated exposed skin. One witness had a name for it: "frack rash." It felt like having alligator skin. At night, children would get intense, sudden nosebleeds; the blood would just pour out. Many of those living in close proximity to a well pad began to become chronically, and inexplicably, sick. (pp. 3–4).

The experiences in Pennsylvania documented in the 2020 grand jury report are consistent with extensive research literature that encompasses the experiences in Pennsylvania as well as elsewhere in the United States. This is clear from a 2019 survey of the literature by Gorski and Schwartz, "Environmental Health Concerns from Unconventional Natural Gas Development." The authors summarize their findings as follows:

> The environmental impacts from UNGD [i.e., fracking] include chemical, physical, and psychosocial hazards as well as more general community impacts. … By 2017, there were a number of important, peer-reviewed studies published in the scientific literature that raised concern about potential ongoing health impacts. These studies have reported associations between proximity to UNGD and pregnancy and birth outcomes; migraine headache, chronic rhinosinusitis, severe fatigue, and

other symptoms; asthma exacerbations; and psychological and stress-related concerns. Beyond its direct health impacts, UNGD may be substantially contributing to climate change (due to fugitive emissions of methane, a powerful greenhouse gas), which has further health impacts. Certain health outcomes, such as cancer and neurodegenerative diseases, cannot yet be studied because insufficient time has passed in most regions since the expansion of UNGD to allow for latency considerations (p. 1).

More recent research, including that focused on health impacts in Pennsylvania, has corroborated the findings from Gorski and Schwartz's 2019 survey. Thus, a 2023 study from the University of Pittsburgh found that children living near gas wells in heavily drilled parts of the state were likelier to develop a relatively rare form of lymphoma, and that nearby residents of all ages had an increased chance of incidents of severe asthma.[6]

## A PENNSYLVANIA EMISSIONS REDUCTION AND CLEAN ENERGY PROGRAM

We describe here a ten-year program through which $CO_2$ emissions in Pennsylvania will fall by approximately 50%. This will enable the state to be in rough alignment with the International Panel on Climate Change's intermediate emissions reduction target of reducing global emissions by 45% as of 2030. As of the most recent 2022 data, overall $CO_2$ emissions in Pennsylvania from burning fossil fuels to produce energy were 214 million metric tons (mmt).[7] A 50% emissions decline would therefore require the level of emissions to be no higher than 107 mmt as of 2034. The state did reduce emissions by a significant 23% between 2005 and 2016, from 281 to 217 mmt. But the state has not achieved further significant cuts since then.

For emissions in the state to fall by 50% over a ten-year period requires that fossil fuel energy consumption in the state will also fall by 50% within ten years. In addition, Pennsylvania exports fossil fuels to other U.S. states as well as, to a lesser extent, other countries. We assume that Pennsylvania's domestic and international fossil fuel export markets will also decline by 50% over this ten-year period. This assumption is consistent with the idea that these other regions will also be moving into alignment with the IPCC's emissions reduction targets.

Based on these assumptions, it follows that over the ten-year period we are considering, production activity and employment in Pennsylvania's fossil fuel industries will decline at the same rate at which energy consumption is declining in Pennsylvania itself, as well as in its export markets—that is, by 50% across the board for all fossil fuel sources. The just transition program will cover the workers in Pennsylvania employed in all of the state's fossil fuel–related sectors, all of which will be phasing down their production activity by 50% over the ten-year period.

## Concurrent Job Creation Through Clean energy Investments

This phase-down of Pennsylvania's fossil fuel production activity will, of course, need to occur in conjunction with the building of a new clean energy infrastructure in the state. The program we developed for Pennsylvania is designed to produce a sufficient clean energy supply in the state that will enable the state to maintain a healthy economic growth path while still reducing its consumption of fossil fuel energy by 50% within ten years. We assume that there will be two areas of large-scale clean energy investments over this ten-year period. They are as follows:

- *Energy efficiency.* Dramatically improving energy efficiency standards in Pennsylvania's stock of buildings, automobiles, public transportation systems, and industrial production processes; and
- *Clean renewable energy.* Dramatically expanding the supply of renewable sources, primarily solar and wind, but also, as supplemental sources, geothermal, small-scale hydro, and low-emissions bioenergy available at competitive prices to all sectors of Pennsylvania's economy.

We estimate that the level of investment needed to achieve Pennsylvania's energy goals will average roughly $26 billion per year (in 2023 dollars) over ten years, with most of the funding being provided by private investors.[8] This estimate assumes that Pennsylvania's economic growth proceeds at an average rate of 1.5% per year. Clean energy investments—from both public and private funding sources combined—will amount to about 2.5% of Pennsylvania's average annual GDP over this ten-year period.

We estimated that investing an average of $26 billion per year in clean energy projects in Pennsylvania over 10 years would generate an average of about 162,000 jobs per year in the state. This includes the three channels through which clean energy investments—as with any investments in all economic activities—generate new employment; that is, the direct, indirect, and induced employment creation channels. The direct channel refers to the jobs created within a sector itself. Mounting solar panels on rooftops is one clear example. The indirect channel refers to jobs created within a given sector's supply chain. An example here is truck drivers delivering solar panels to a construction site. Induced jobs are those generated by multiplier effects—that is, the job creation generated by increases in consumption spending resulting from the increased incomes provided by newly created jobs.

We estimate that of the 162,000 jobs generated by $26 billion in annual clean energy investments in Pennsylvania, about 107,000 will be generated through either direct or indirect channels within the various clean energy investment projects. This amounts to about 1.6% of Pennsylvania's overall 2024 labor force of 6.5 million people. The remaining 55,000 jobs will be induced through the increased overall spending levels in the state resulting from direct and indirect growth in employment. The 162,000 in overall employment expansion through Pennsylvania's clean energy investments would represent about 2.5% of the state's workforce.

Focusing on the direct employment channel, the clean energy investments will produce new job opportunities in a wide range of areas, including construction, sales, management, production, engineering, and office support. As of the 2023 Pennsylvania labor market, the average pay in most sectors ranges between about $80,000 and $95,000. The weighted average pay level for the full set of clean energy–related jobs in Pennsylvania is $81,033. This includes jobs in retrofitting buildings, industrial efficiency, electrical grid upgrades, and building out the state's solar and wind energy sectors. The one area where pay is significantly lower is mass transit, where the average pay is about $40,000. Average pay is highest in industrial efficiency, at about $108,000.[9]

## CLEAN ENERGY INVESTMENTS IN PENNSYLVANIA THROUGH BIDEN-ERA PROGRAMS

The enactment under the Biden administration of the BIL in November 2021 and the IRA in August 2022 created a major new pool of available funding for advancing a viable clean energy transition program in the United States overall, and in Pennsylvania specifically. The BIL consists of a wide range of public programs in the areas of manufacturing, and infrastructure, as well as clean energy, with a total annual budget for the U.S. economy overall of $170 billion per year over five years. Roughly one-third of this total funding level, or roughly $60 billion per year—is designated for programs that, in broad terms, support energy efficiency and renewable energy projects.[10]

The IRA also includes a wide range of specific programs, again with about one-third of these projects broadly aimed at supporting energy efficiency and renewable energy investments. But funding through the IRA is distinct in that it mostly consists of subsidies for private investments, either in the form of tax credits or loan guarantees. Moreover, with a high proportion of individual IRA programs, the availability of subsidy support is uncapped—that is, support for individual private investment projects is available without specified limits in the aggregate. As such, the total amount of clean energy investment financing that could result from the IRA could potentially reach very high levels, depending on the take-up rate by private investors and consumers. A midrange estimate for IRA-funded clean energy for the overall U.S. economy is about $30 billion per year over ten years. But credible higher-end estimates are three to four times higher, ranging up to $120 billion per year over a decade.[11]

Of course, as of this writing in January 2025, it is uncertain as to how much of the BIL and IRA will survive under the Trump administration. Consistent with his general climate denialism, Trump has never expressed support for any clean energy investment programs. But he could potentially alter his position during his second presidential term, given the benefits that have resulted to date, and would be ongoing, from these programs in terms of job creation and community development.

For our present discussion, our focus is to estimate the extent of funding that the IRA and BIL have provided to date in Pennsylvania and the prospects for further support if these programs are maintained going forward. According to a website that was maintained by the Biden White House, Invest.gov, total funding commitments in Pennsylvania since the IRA and BIL programs were enacted, including both private and public investments, amounted to about $18 billion. Of this total funding level, roughly half of the reported total—that is, about $9 billion, could be categorized as clean energy investments. Of this $9 billion in funding announcements, about $7 billion are private investment projects subsidized in part by IRA funds. The remaining $2 billion are public sector projects supported by both BIL and IRA programs. From the information provided at Invest.gov, it is not possible to establish with confidence the expected time frame over which these $9 billion in clean energy investment commitments will flow into the Pennsylvania economy. As a rough approximation, we estimate that, on average, spending per project will occur over a three-year period. This suggests that approximately $3 billion per year would be spent in Pennsylvania on clean energy investments through projects that have been announced to date at Invest.gov. This level of annual clean energy investment spending is, of course, significant. But it is also not close to sufficient relative to the $26 billion per year in spending that we estimate is necessary for Pennsylvania to build out a clean energy–dominant infrastructure within ten years. This, again, is the clean energy infrastructure that would be capable within ten years of providing a sufficient supply of clean energy in Pennsylvania that would substitute for the 50% reduction in fossil fuel energy consumption in the state.

A major challenge for Pennsylvania, as elsewhere, is, therefore, how to dramatically scale up clean energy investments in the state. Addressing this question is beyond the scope of this paper.[12] For our present discussion, what is most critical is to be able to understand the scale at which clean energy investments in Pennsylvania will be needed, and how this project, operating at this requisite scale, will expand new job opportunities that are more than sufficient to offset the job losses that will result through the phase-out of fossil fuel production in the state.

As another rough approximation, derived from our estimate that $26 billion per year in clean energy investments would generate about 160,000 new jobs in Pennsylvania, it follows that the roughly $3 billion per year allocated to date for Pennsylvania through the IRA and BIL would generate about 20,000 jobs per year in the state.[13]

## FOSSIL FUEL PHASE-OUT AND JUST TRANSITION

The issue on which we focus in this section is what the impact will be on workers employed in Pennsylvania's fossil fuel–based industries. We assume that, over the ten-year period (2026–2035), production activity and employment in all of Pennsylvania's fossil fuel–based industries will decline at approximately the same 50% rate as energy consumption in the state. In particular, we develop here a just transition program for the workers in these fossil fuel sectors who will face displacement as a result of this contraction in the state's fossil fuel–based sectors.

Our primary focus in this section is on the *direct* jobs that will be lost in Pennsylvania through the contraction of the state's fossil fuel–based industries. The workers currently employed in these jobs will be the ones who will be most in need of just transition support as Pennsylvania phases out these $CO_2$-generating activities. The jobs that will be lost through the indirect and induced channels will be more diffuse in their characteristics. A high proportion of the jobs lost through indirect channels are likely to match up reasonably well with those in the clean energy economy, including in areas such as administration, clerical, professional services, and transportation services. The characteristics of the induced jobs created will simply reflect the overall characteristics of Pennsylvania's present-day workforce. The job losses that will result through the indirect and induced channels can therefore be appropriately managed through the same set of policies that are available to all workers in Pennsylvania who experience unemployment.

## Measuring Direct Employment Levels

In Table 1, we show employment levels for the 14 fossil fuel and ancillary industries in Pennsylvania as of 2023. As of 2023, there are 48,293 people employed in the fossil fuel and ancillary industries in Pennsylvania. Of these, 12,375 (26%) are employed in oil and gas extraction, 8,416 (17%) work in oil and gas support activities, and 5,306 work in natural gas distribution (11%). Thus, these three sectors—extraction, support activities, and natural gas distribution—together account for 54% of total employment in all of Pennsylvania's fossil fuel–based industries. The other major employment category is coal mining, with 7,292 jobs at 15% of the total.

## Characteristics of Fossil Fuel–Based Industry Jobs

Table 2 provides basic figures on the characteristics of the direct jobs in Pennsylvania for workers in fossil fuel–based sectors. We first see that, on average, these are relatively high-paying jobs. The average overall compensation is roughly $118,000. This is nearly 50% more than the $81,033 average pay level for jobs in Pennsylvania's various clean energy sectors. It is also about 9% more than the $108,000 average pay level for workers in the industrial efficiency sector, who are, on average, the highest paid workers employed in Pennsylvania's clean energy sectors.

In terms of private health insurance coverage, the fossil fuel industries are, for the most part, providing coverage for their workers, with about 75% of workers receiving employer-based insurance. This level of health insurance coverage is significantly higher than the 60.4% of U.S. workers overall who receive health insurance through their employer.

Union membership is at about 15% for Pennsylvania's fossil fuel–based workers. This is much higher than the figure for the overall U.S. private sector, of 5.9%.[14]

Table 2 also reports figures on educational credential levels for workers in the fossil fuel–based sectors, as well as the percentages of workers who are women and people of color. With respect to educational credentials, the overall level of attainment is relatively high, with about 40% having a bachelor's degree or higher, and another 22% having have some college or an associate degree. The remaining 38% have high school diplomas or less. Women account for only 19% of the workforce, and non-White workers account for roughly 14% of the total workforce.

Table 1. Number of Workers in Pennsylvania
Employed in Fossil Fuel–Based Industries, 2023

| Industry | 2023 employment levels | Industry share of total fossil fuel–based employment |
| --- | --- | --- |
| *Fossil fuel industry total* | *48,293* | *100.0%* |
| Oil and gas extraction | 12,375 | 25.6% |
| Support activities for oil/gas | 8,416 | 17.4% |
| Coal mining | 7,292 | 15.1% |
| Natural gas distribution | 5,306 | 11.0% |
| Wholesale—petroleum and petroleum products | 3,447 | 7.1% |
| Fossil fuel electric power generation | 2,647 | 5.5% |
| Pipeline transport | 2,542 | 5.3% |
| All other petroleum and coal products manufacturing | 1,729 | 3.6% |
| Drilling oil and gas wells | 1,273 | 2.6% |
| Pipeline construction | 1,123 | 2.3% |
| Petroleum refining | 1,107 | 2.3% |
| Mining machinery and equipment manufacturing | 600 | 1.2% |
| Support activities for coal | 311 | 0.6% |
| Oil and gas field machinery and equipment manufacturing | 125 | 0.3% |
| **Total fossil fuel employment as share of Pennsylvania employment** *(Pennsylvania 2023 employment = 6,295,653)* | 0.8% | |

Source: IMPLAN (https://implan.com); Pennsylvania employment is from the BLS Local Area Unemployment Statistics program (https://www.bls.gov/lau).

In Table 3, we gain further detailed information on workforce and employment conditions for workers in Pennsylvania's fossil fuel–based industries. We show the most prevalent job categories and the representative occupations in each job category.

The key finding that emerges from these tables is that Pennsylvania's fossil fuel–based industries provide a wide range of employment opportunities for the nearly

Table 2. Characteristics of Workers Employed
in Pennsylvania's Fossil Fuel-Based Sectors

|  | Fossil fuel–based industries |
| --- | --- |
| Average total compensation (2023 dollars) | $118,083 |
| Health insurance coverage | 74.7% |
| Retirement benefits | 54.2% |
| Union membership coverage | 14.6% |
| **Educational credentials** | |
| Share with high school degree or less | 37.7% |
| Share with some college or associate degree | 22.4% |
| Share with bachelor's degree or higher | 40.0% |
| **Racial and gender composition of workforce** | |
| Percentage of non-White workers | 13.6% |
| Percentage of female workers | 19.0% |

Source: IMPLAN; Current Population Survey data files 2022–2024. See appendix for details.

50,000 workers currently employed in these industries. As we see, the largest share of jobs, at roughly 18%, are in management. The next largest category, at roughly 16%, is "transportation and material movers." Office and administrative support is the next largest category, at about 15%, followed by market research and human resources, at about 9%. In combination, these four job categories account for nearly 60% of total employment in Pennsylvania's fossil fuel–based industries. It is important to note that with this 60% figure for total fossil fuel–based employment, most of the occupations in these job categories—in management, transportation, office support, and business operations—require skills that are not specific to the activities of the fossil fuel industries per se. Rather, these job categories mostly require skills that are transferable to other economic sectors.

With the other two fossil fuel–based job categories employing more than 5% of workers shown in Table 3—that is, extraction at 8.6% and production at 6.4%, respectively, of total employment—the skill requirements will be more specific to the fossil fuel activities themselves. For this roughly 15% of Pennsylvania's fossil fuel–based industries' workforce, the challenge will therefore be greater to transition these workers into new employment situations as their fossil fuel–based jobs are phased out. More generally, what these employment figures underscore is that any just transition program to support displaced workers in Pennsylvania's fossil fuel–related industries will need to be focused on the specific background and skills of each of the impacted workers.

Table 3. Prevalent Job Types in Pennsylvania's Fossil Fuel–Based Industries
(Job Categories with 5% or More Employment)

| Job category | Percentage of direct jobs lost | Representative occupations |
|---|---|---|
| Management | 18.3% | Marketing managers; computer and information systems managers; general and operations managers |
| Transportation and material movers | 16.4% | Supervisors of transportation and material moving workers; pumping station operators; hand laborers and freight, stock, and material movers |
| Office and administrative support | 15.1% | Customer service representatives; bookkeeping, accounting, and auditing clerks; secretaries |
| Business operations specialists | 9.1% | Market research analysts and marketing specialists; project management specialists; human resource workers |
| Extraction | 8.6% | Derrick, rotary drill, and service unit operators in oil, gas, and mining; earth drillers; explosives workers, ordnance handling experts, and blasters |
| Production | 6.4% | Plant and system operators, first-line supervisors of production and operating workers; inspectors |

Source: Current Population Survey data files 2022–2024.
Note: Due to small sample sizes, these estimates are based on the mid-Atlantic region, which includes Pennsylvania, New York, and New Jersey, rather than Pennsylvania only.

## Features of a Just Transition Program

We outline here a just transition program for workers who face job losses through direct channels from the 50% contraction of the state's fossil fuel industries. The program has three major elements. These are:

- Guaranteeing pensions for workers in affected industries who will retire up until the year 2035

- Guaranteeing re-employment for workers facing displacement
- Providing income, retraining, and relocation support for workers facing displacement

We describe each feature of this program in what follows and provide estimates of the costs of effectively operating each measure within the overall program. The detailed policy package includes five components. These are:

- Pension guarantees for retired workers who are covered by employer-financed pensions, starting at age 65
- Re-employment for displaced workers through an employment guarantee, with 100% wage insurance. With wage insurance, workers are guaranteed that their total compensation in their new job will be supplemented to reduce any losses relative to the compensation they received working in the fossil fuel–based industry
- Retraining, as needed, to assist displaced workers in obtaining the skills required for a new job
- Relocation support for 50% of displaced workers, assuming only 50% will need to relocate
- Full just transition support for workers 65 and over who choose not to retire

## Steady Versus Episodic Industry Contraction

Before presenting the cost estimate calculations, it is critical to note how any such policy measures will be affected by the conditions under which the fossil fuel–based industry contraction occurs in Pennsylvania. Specifically, the scope and cost of any set of just transition policies will depend substantially on whether the contraction is steady or episodic.

Under a pattern of steady contraction, there will be uniform annual employment losses over the ten-year period in the affected industries. But it is not realistic to assume that the pattern of industry contraction will necessarily proceed at a steady rate. An alternative pattern would entail relatively large episodes of employment contraction, followed by periods in which no further employment losses are experienced. This type of pattern would occur if, for example, one or more relatively large firms were to undergo large-scale cutbacks at one point in time as the industry overall contracts, or even for such firms to shut down altogether.

The costs of a ten-year just transition will be much lower if the transition is able to proceed smoothly rather than through a series of episodes. One reason is that, under a smooth transition, the proportion of workers who will retire voluntarily in any given year will be substantially greater than if several large businesses were to shut down abruptly and lay off their full workforce at a given point in time. Another factor is that it will be easier to find new jobs for displaced workers if the pool of such workers at any given time is smaller.

Table 4. Attrition by Retirement and Job Displacement
for Fossil Fuel Workers in Pennsylvania

|  | Fossil fuel workers |
| --- | --- |
| 1. Total workforce as of 2023 | 48,293 |
| 2. Job losses over 10-year transition, 20262035 | 24,147 |
| 3. Average annual job loss over 10-year production decline (= row 2/10) | 2,415 |
| 4. Number of workers reaching 65 over 2026–2035 (= row 1 × % of workers 55 and over in 2025) | 9,176 *(20.9% of all workers)* |
| 5. Number of workers per year reaching 65 during 10-year transition period (= row 4/10) | 918 |
| 6. Number of workers per year retiring voluntarily | 734 *(80% of 65+ workers)* |
| 7. Number of workers requiring re-employment (= row 3 – row 6) | 1,681 |

Source: The 80% retirement rate for workers over 65 derived from U.S. Bureau of Labor Statistics (https://www.bls.gov/cps/cpsaat03.htm). According to these BLS data, 20% of 65+ year-olds remain in the workforce.

We proceed here by assuming that Pennsylvania will successfully implement a relatively steady contraction of its fossil fuel sectors. This should be realistic, as long as the relevant policy makers remain focused on that goal.

## Estimating Attrition by Retirement and Job Displacement Rates

In Table 4, we show figures on annual employment reductions in Pennsylvania's fossil fuel–based industries over 2026–2035 that would result from a steady contraction of these industries.

We also show the proportion of workers who will move into voluntary retirement at age 65 by 2035. Once we know the share of workers who will move into voluntary retirement at age 65, we can then estimate the number of workers who will be displaced through the 50% contraction of fossil fuel production in the state. As described above, the just transition program will provide support for all displaced workers through a re-employment guarantee along with wage insurance, retraining, and relocation support.

All forms of just transition support will also be fully available to workers 65 and over who choose to continue working. We therefore need to estimate how many workers 65 and older are likely to choose to remain employed. For the fossil fuel sector taken as a whole, we approximate that about 20% of workers who are 65 and over choose to continue on their jobs.[15] We therefore assume that this same 20% of older workers will choose to continue working while the fossil fuel–based sectors undergo their contractions between 2026 and 2035. Specifically, we incorporate into our calculations in Table 4 an estimate that of the total number of workers reaching age 65 in any given year, 80% will retire voluntarily while 20% will choose to continue working.

We can see, step by step, how these various considerations come into play through the figures we show in Table 4. As we again see in column 2 of Table 4, as of the most recent 2023 figures, 48,293 workers in Pennsylvania were employed in all fossil fuel–based industries. Given the 50% contraction in all fossil fuel–based industries in the state, this means that total employment in these sectors will fall by 24,147 as of 2035. It therefore also means that the same number of jobs, 24,147, will be retained. If we then assume that the contraction in these industries proceeds at a steady rate between 2026 and 2035, this means that 2,415 jobs in these industries will be lost each year, as we see in row 3 (i.e., 24,147 job losses in total/ten years of industry contraction = 2,415 job losses per year).

We see in row 4 that, of the workers presently employed in these sectors in Pennsylvania, 9,176, or 21%, will be between 55 and 65 years old over 2026–2035. If all these workers were to voluntarily retire at a steady rate over 2026–2035, this would mean that 918 workers will move into retirement every year over the ten-year period. However, we are assuming that only 80% of these workers will retire once they reach 65. That is, as we see in row 6, we estimate that 734 workers employed in these sectors will retire voluntarily every year between 2026 and 2035.

Given that total job losses each year will average 2,415 over the 2026–2035 period, that in turn means that the total number of workers currently employed in Pennsylvania's fossil fuel–based sectors that will require re-employment will be 1,681 per year. We show this figure in row 7 of Table 4.

This is a critical result. The immediate point it establishes is that the just transition program will need to focus on two areas: (1) guaranteeing the pensions for the 734 workers per year moving into voluntary retirement; and (2) providing all the forms of re-employment support, including the re-employment guarantee, for the 1,681 workers per year facing displacement. Of course, these figures are not meant to be understood as precise estimates but rather to provide broadly accurate magnitudes. Among other factors beyond what these figures themselves show, we again have to recognize that the pattern of contraction is not likely to be as steady as is being assumed in our calculations.

Nevertheless, precise details aside, it is the overall finding that these results firmly establish that is most central: that the number of workers in Pennsylvania who are likely to experience job displacement through the state's transitioning away from $CO_2$-generating energy sources will be small—indeed, the number of workers facing displacement, if not exactly 1,681 per year, should be, under most circumstances, below 2,000 per year.

## Cost Estimates for a Just Transition Program
*Pension Guarantees for Retiring Workers*

What becomes clear from the evidence on the steady rate of contraction for Pennsylvania's fossil fuel–related industries is that guaranteeing workers' pension funds must be a centerpiece of the state's overall just transition program. This is especially important, given that the fossil fuel–based enterprises will likely face major financial challenges through experiencing sharp contractions between 2026 and 2035. Under these circumstances, these firms may not consider their pension fund commitments to be a top financial priority. Despite this, guaranteeing workers' pensions as a first-tier financial obligation for employers can be established through regulatory policies. For example, the State of Pennsylvania could work in coordination with federal regulators at the Pension Benefit Guarantee Corporation to place liens on company assets when pension funds are underfunded. Through such measures, the pension funds for most of the affected workers can be protected through regulatory intervention alone, without the government having to provide financial infusions to sustain the funds.[16]

*Guaranteed Re-Employment*

New employment opportunities will certainly open up in the expanding clean energy sectors, with approximately 107,000 new direct plus indirect jobs created per year in Pennsylvania through clean energy investments at the level of $26 billion per year (see Table 2.16). An additional 55,000 jobs will also likely be generated through the induced job creation channel—that is, multiplier effects resulting from the 107,000 new jobs generated through the direct and indirect employment channels. A high proportion of new state clean energy projects is likely to be financed at least partially through public sector funding. Given such public sector funding, the state could require job preference provisions for displaced workers. Again, our estimate of the number of displaced workers per year who will need re-employment is about 1,700 in total. It will not be difficult for the state to set aside 1,700 to 2,000 guaranteed jobs for these displaced workers, or, for that matter, even, say, 10,000 jobs, as needed for this purpose.

This remains true even if we assume that the level of clean energy investments in Pennsylvania was much more modest than the $26 billion per year figure we have projected as necessary for achieving the 50% $CO_2$ reduction target by 2035. Thus, as we discussed above, the level of IRA/BIL-supported investments for Pennsylvania had totaled about $9 billion by the end of the Biden administration. We roughly assumed above that this $9 billion total would be allocated over three years at $3 billion per year. That level of clean energy investments, while not close to adequate for sufficiently expanding the state's clean energy infrastructure, would nevertheless generate about 20,000 new jobs in Pennsylvania. This level of job creation would be approximately ten times more than the number of workers who would be displaced through the state's 50% fossil fuel phase-out through 2035.

*Income Support Through Wage Insurance*

Though it will not be difficult to find new employment opportunities for the roughly 1,700 fossil fuel–based workers who will be displaced annually on average, there is a high likelihood that for workers currently employed in fossil fuel–based industries and re-employed in clean energy activities, their new jobs will be at lower pay levels than their previous jobs. As we have seen, the average compensation for fossil fuel–based workers in Pennsylvania at present is $118,083. This compares with the average compensation in the clean energy areas, ranging for the most part, as noted above, between about $80,000 and $95,000 in the various specific sectors. The average weighted compensation figure for the full set of direct jobs generated by clean energy investments is, again, $81,035. It will therefore be necessary for the fossil fuel–based sector workers to be provided with wage insurance so that they experience no income losses in their transition from fossil fuel industry jobs to new positions.

To provide some initial specifics on the costs of providing wage insurance for displaced workers who move into jobs at lower pay levels, we propose that all displaced workers facing pay cuts receive 100% compensation insurance for three years. That is, they will be paid the full difference between any disparities in the compensation they receive in their new jobs relative to what they received in their previous jobs in fossil fuel–related industries.

Table 5 presents a framework for calculating a rough estimate of the costs for such a compensation insurance program. In row 1, the table shows the figures we have seen in Table 4 on the number of displaced workers in the fossil fuel–based sectors— that is, 1,681 workers per year. Row 2 then shows their average compensation level of $118,083. In row 3, we show the weighted average compensation level for all of Pennsylvania's clean energy sectors, which, as noted above, is $81,033. From this difference in average compensation levels, we then calculate that the annual cost of compensation insurance for 1,681 workers will be about $62 million.

*Retraining Support*

As we have discussed above, the range of new jobs that are being generated through clean energy investments varies widely in terms of their formal educational credentials as well as special skill requirements. A majority of the jobs will require skills closely aligned with those that the displaced workers used in their former fossil fuel–based industry jobs. These include most management, administrative, and transportation-related positions throughout the clean energy industries. In other cases, new skills will have to be acquired to be effective in the clean energy industry jobs. For example, installing solar panels is obviously distinct from laying oil and gas pipelines. This is why a just transition program must include a provision for retraining for displaced fossil fuel–based industry workers whose skills do not transfer readily into the state's other areas of employment, in clean energy or otherwise. For this discussion, we assume, as a high-end figure, that 50% of the 1,681 displaced workers per year—that is, 840 workers per year—will need access to significant retraining opportunities after experiencing displacement from their fossil fuel industry–based jobs.

Table 5. Estimating Costs of 100% Compensation Insurance for Displaced
Workers in Pennsylvania's Fossil Fuel–Based Sectors

| | |
|---|---|
| 1. Number of fossil fuel–based displaced workers per year requiring re-employment | 1,681 |
| 2. Average compensation for displaced workers (2023 dollars) | $118,083 |
| 3. Average compensation for clean energy sector jobs (2023 dollars) | $81,033 |
| 4. Average compensation difference between fossil fuel–based and clean energy jobs (= row 2 – row 3) | $37,050 |
| 5. Annual cost of compensation insurance for 1,681 workers (= row 4 x row 1) | $62.3 million |
| 6. Total cost of compensation insurance for 3 years (= row 5 x 3) | $186.8 million |

Source: See Tables 2 and 4.

We envision two components of this job retraining program for these 840 displaced workers. The first will be to finance the actual training programs themselves. We can estimate this with reference to the overall costs of providing community college education. The average figure for in-state tuition for community college in Pennsylvania is around $11,000.[17] We then also allow an additional $3,000 per year per worker to cover other expenses during their training program, such as purchases of textbooks and equipment. We assume that workers would require the equivalent of two full years of training, which they would most likely spread out on a part-time basis as they move into their guaranteed jobs. By this measure, the average cost of the training program for 840 workers would be about $12 million per year.

### Relocation Support

Some displaced workers will need to be relocated to begin their new jobs. For the purposes of our discussion, we again assume that half of the 1,681 displaced workers per year will need relocation allowances, at an average of $75,000 per displaced worker.[18] That would bring the annual relocation budget to about $63 million for 840 workers each year.

### Overall Costs for Supporting Displaced Workers

In Table 6, we show estimates of the full costs of providing this set of wage insurance, retraining, and relocation support for 1,681 workers per year. As Table 6 shows, the total level of annual spending will vary, depending largely on the number of cohorts of displaced workers receiving just transition benefits.

For example, in 2026, the first cohort of 1,681 displaced workers will receive support through the just transition program, including wage insurance, retraining,

Table 6. Total and Annual Average Costs for Just Transition Support for Displaced Fossil Fuel–Based Workers in Pennsylvania, 2026–2035

| Year | Income support (3 years of support for 1,681 workers) | Retraining support (2 years of support for 840 workers) | Relocation support (1 year of support for 840 workers) | Total (columns 1 + 2 + 3) |
|---|---|---|---|---|
| 2026 | $62.3 million | $12 million | $63.0 million | $137.3 million |
| | (1 cohort) | (1 cohort) | | |
| 2027 | $124.6 million | $24 million | $63.0 million | $211.6 million |
| | (2 cohorts) | (2 cohorts) | | |
| 2028 | $186.9 million | $24 million | $63.0 million | $273.9 million |
| | (3 cohorts) | (2 cohorts) | | |
| 2029 | $186.9 million | $24 million | $63.0 million | $273.9 million |
| | (3 cohorts) | (2 cohorts) | | |
| 2030 | $186.9 million | $24 million | $63.0 million | $273.9 million |
| | (3 cohorts) | (2 cohorts) | | |
| 2031 | $186.9 million | $24 million | $63.0 million | $273.9 million |
| | (3 cohorts) | (2 cohorts) | | |
| 2032 | $186.9 million | $24 million | $63.0 million | $273.9 million |
| | (3 cohorts) | (2 cohorts) | | |
| 2033 | $186.9 million | $24 million | $63.0 million | $273.9 million |
| | (3 cohorts) | (2 cohorts) | | |
| 2034 | $186.9 million | $24 million | $63.0 million | $273.9 million |
| | (3 cohorts) | (2 cohorts) | | |
| 2035 | $186.9 million | $24 million | $63.0 million | $273.9 million |
| | (3 cohorts) | (2 cohorts) | | |
| 2036 | $124.6 million | $12 million | — | $136.6 million |
| | (2 cohorts) | (1 cohort) | — | |
| 2037 | $62.3 million | — | — | $62.3 million |
| | (1 cohort) | — | — | |
| Total | $1.9 billion | $240 million | $630 million | $2.7 billion |
| Average annual costs | $155.7 million (12 years of support) | $21.8 million (11 years of support) | $63.0 million (10 years of support) | $240.5 million (12 years of support) |

Source: See Tables 4 and 5. All figures are in 2023 dollars.

and relocation support, as needed. As we can see in column 4, these full costs will amount to $137.3 million in 2026. Costs increased in 2027, since we now have two cohorts of displaced workers receiving income and retraining support as well as one cohort receiving relocation support. Thus, total costs in 2026 rise to $211.6 million. In 2027, there are now three cohorts of displaced workers receiving income support, along with two cohorts receiving retraining support and, again, one cohort receiving relocation support. This totals to $273.9 million, the figure that then prevails through

2036. In 2036 and 2037, with smaller cohorts eligible for income and retraining support, and no further cohorts receiving relocation support, the costs of the program fall correspondingly to $136.6 million, then to $62.3 million.

In total, just transition benefits provided to 1,683 displaced workers per year in Pennsylvania will total to $2.7 billion, or an average of $240.5 million per year over 12 years, in total costs and about $167,000 per worker.

*Transitional Support for Workers Facing Indirect and Induced Job Losses*

It should not be a challenge, either administratively or financially, to provide transition support for the relatively small number of workers facing displacement through indirect and induced job channels. This is especially the case because, on balance, there should be no jobs lost in Pennsylvania through the induced employment channel after we take into account the just transition program for workers who experience displacement through the direct employment channel. This is because, as we have described above, induced employment effects refer to the expansion of employment that results when people in any given industry—such as clean energy or fossil fuels— spend money and buy products. This increases overall demand in the economy, which means that more people are hired into jobs to meet this increased demand. It follows that the loss of incomes through a contraction of employment will create a reverse-induced employment effect. People will have less money to spend, overall demand for goods and services will contract, and therefore the demand for employees will decline correspondingly. However, our proposed just transition program provides that workers facing displacement through the direct jobs channel will be guaranteed re-employment at a compensation level equal to what they were earning before they became displaced. It follows that implementing the just transition program will mean that there will also be no reverse-induced employment effects in Pennsylvania even as the fossil fuel–based industries themselves contract.

## CONCLUSION

The extraction of natural gas from the Pennsylvania expanse of the Marcellus Shale deposit through fracking operations, which began in 2008, has delivered some economic benefits to working people and communities in Pennsylvania. As we have seen, at their respective peak levels in 2014 and 2017, roughly 24,000 people in Pennsylvania were employed in fracking operations, and average pay reached $127,000. At the same time, fracking in Pennsylvania has generated severe public health and environmental impacts, including groundwater contamination, and a range of chemical, physical, and psychological hazards.

The negative impacts of fracking operations in Pennsylvania extend well beyond the state's borders. This is because burning fossil fuels to produce energy—including natural gas extracted from the Earth through fracking operations—is, by far, the most important driver generating the global climate crisis. Phasing out all fossil fuel consumption is therefore a first-order priority for moving onto a viable climate stabilization path in Pennsylvania and throughout the global economy.

At the same time, people still need to consume energy to light, heat, and cool buildings; to power cars, buses, trains, and airplanes; and to operate computers and industrial machinery, among other uses. As such, making progress toward climate stabilization requires building, throughout the global economy, a new energy infrastructure whose foundations are high-efficiency and clean renewable energy sources.

As we have reviewed, the project of building a clean energy infrastructure in Pennsylvania, capable of providing the state with roughly half of its energy needs as of 2035 while fossil fuel consumption declines by a corresponding 50%, would generate far more jobs in the state than have been created through fracking. By our estimate, investing about $26 billion per year in clean energy investments in the state, roughly equal to 2.5% of the state's GDP, would deliver sufficient energy to supply the state with 50% of its energy needs. These clean energy investments would generate roughly 160,000 jobs in the state in a wide range of occupations.

Nevertheless, it is still the case that the phase-out of fracking and other fossil fuel production will entail job losses and displacement for workers that are now employed in the fossil fuel industry. As we show, after taking into account voluntary retirements, the number of fossil fuel sector-based workers who would face displacement would be roughly 1,700 per year though the ten-year phase down of the state's fossil fuel production activity between 2026 and 2035.

It will not be a major challenge to find new jobs for these 1,700 workers per year, given the 160,000 jobs that will be generated concurrently through clean energy investments in the state, as well as, more generally, an overall employment level in Pennsylvania of 6.2 million people. The 1,700 displaced workers per year would thus constitute 1.1% of the jobs generated by Pennsylvania's clean energy investments and 0.03% of overall employment in the state.

Nevertheless, these 1,700 workers per year facing displacement deserve generous transition support policies—what we have termed, following Tony Mazzochi, a just transition. These just transition policies should include pension, employment, and income guarantees, in addition to, as needed, retraining, and relocation support. We estimated that a just transition program for these workers, including wage insurance, retraining, and relocation programs, would cost in the range of $240 million per year. This amounts to about 0.02% of Pennsylvania's current GDP.

In short, our paper shows how, between 2026 and 2035, Pennsylvania could phase out 50% of all its fossil fuel production activities—including its fracking operations— while also providing generous support for working people to transition out of their fossil fuel industry jobs and into activities that both raise public health and environmental standards in the state and contribute significantly toward a viable global climate stabilization project.

# APPENDIX

## Estimating Worker Characteristics

Our estimates of worker characteristics, presented in Tables 2 through 4, are based on data from the U.S. Labor Department household survey, the Current Population Survey (CPS), administered by the U.S. Census Bureau for the Bureau of Labor Statistics (see www.bls.gov/cps). For a full discussion of the measures we use and how we identify workers in the fossil fuel sectors using IMPLAN and CPS combined, see Appendix 2 of Pollin et al. (2021). For this chapter, we use the most up-to-date, post-COVID recovery CPS data files available (i.e., 2022–2024) as well as the 2023 IMPLAN data on fossil fuel sector employment. The one exception to this is that, as we discuss below, we use different data sources for our compensation figures.

As with the Pollin et al. (2021) study, we need to pool CPS data across years (2022–2024) in order to achieve adequate sample sizes and generate reasonable estimates of worker characteristics. We also need to pool, in most cases, across geographic units beyond Pennsylvania. Except for our estimates of the percentage of fossil fuel sector workers who are age 55 years old and over in 2024 (Table 4) and the compensation figures, these estimates are based on the mid-Atlantic region, which includes Pennsylvania, New York, and New Jersey, rather than Pennsylvania only.

## Compensation Estimates

The compensation figures for fossil fuel sector jobs in this chapter are based on the U.S. Labor Department's Quarterly Census on Employment and Wages (QCEW). The QCEW is the BLS' establishment-based survey and is a near census of all jobs in the U.S., with estimates available at the national, state, MSA, and county levels (see https://www.bls.gov/cew). This departs from our method of using IMPLAN compensation figures in our 2021 report. We changed our data source for our fossil fuel sector earnings estimates due to the unusual—and implausible—volatility in the IMPLAN compensation data for Pennsylvania's oil and gas extraction industry from 2018 through 2023. The oil and gas extraction industry comprises a quarter of Pennsylvania's fossil fuel sector jobs (see Table 1), so any volatility in the data for this sector significantly impacts our measures for the overall fossil fuel sector in Pennsylvania.

Specifically, IMPLAN provides employment and compensation data for wage and salary workers, as well as proprietors (see Appendix 2 of Pollin et al. 2021 for discussion as IMPLAN data). We show in Table A.1 the average annual compensation figures of oil and gas extraction industry wage/salary workers and proprietors separately (columns 1 and 2) and then combined (column 3). Wage and salary workers make up between 2% and 33% of total employment in this sector.

The IMPLAN estimates indicate that wage and salary workers' earnings more than doubled between 2018 and 2023. This contrasts sharply with the wage trends indicated by the BLS' QCEW data (column 4; also see Figure 1). The BLS QCEW data indicate that wage and salary workers' earnings rose between 2018 and 2019 and then fell below 2018 levels in 2023. Although the QCEW data are for wages only, whereas the IMPLAN

Table A.1. Compensation Estimates for the Oil and
Gas Extraction Industry (2023 Dollars), 2018–2023

| Year | IMPLAN: Average annual compensation | | | BLS QCEW: Average annual wages |
| | Wage/salary employees | Proprietors | Wage/salary employees and proprietors, combined | Wage/salary employees |
| --- | --- | --- | --- | --- |
| 2018 | $196,701 | $28,163 | $52,119 | $148,903 |
| 2019 | $260,197 | $179,687 | $168,932 | $154,574 |
| 2023 | $456,171 | $65,620 | $160,213 | $144,116 |

data include wages and benefits, these trends should nevertheless be similar, as they are both measures of compensation for wage-earning workers.[19]

The IMPLAN proprietor compensation figures (column 2) are even more volatile than the IMPLAN wage and salary figures. However, this may be explained by the fact that proprietors' compensation reflects earnings from self-employment or other types of business income and therefore includes both earnings gains and losses. Additionally, these figures may reflect other year-to-year accounting differences that are not applicable to wages.

It is unclear what the source of these implausible trends is in IMPLAN earnings, particularly for wage and salary earners. As a result, we do not view the IMPLAN compensation estimates as reliable for our current analysis. Instead, we use the average annual wage estimates from the BLS' QCEW. Moreover, we use the average data from 2019 (just prior to the COVID pandemic) and 2023 to smooth out the shock to the economy of the COVID pandemic that may linger in the 2023 data and thereby better reflect the conditions of today's fully recovered economy.

We do not observe the same type of volatility in the IMPLAN earnings data for clean energy–sector jobs. Therefore, we continue to use IMPLAN data for the clean energy–sector compensation estimates and use the average of the 2019 and 2023 figures. We present these figures in Table A.2.

## ENDNOTES

1. https://tinyurl.com/3pb4kn63

2. This study summarizes and updates the main findings of Pollin et al. (2021). All statistical results that are not explicitly derived in this paper itself are presented in the 2021 Pollin et al. study.

3. See, e.g., Chomsky and Pollin (2020) for a general overview of Green New Deal policy perspectives and Pollin (2023) for a more formal presentation of this framework.

4. https://www.eia.gov/state/seds

5. Office of the Attorney General, Commonwealth of Pennsylvania (2020).

6. Buchanich et al. (2023)

7. An additional source of CO2 emissions in Pennsylvania, as elsewhere, is generated through combusting wood, plants, and waste materials to produce bioenergy. This additional source of CO2 emissions from consuming bioenergy constitutes a negligible share of the state's overall CO2 emissions. For simplicity, we therefore do not include bioenergy emissions in our estimates of total CO2 emissions in the state, or in our calculations for moving Pennsylvania onto a zero emissions trajectory. https://www.eia.gov/state/seds

8. This $26 billion per year investment figure is expressed in 2023 dollars, whereas, in our 2021 study (Pollin et al. 2021), the overall clean energy investment figure we reported was $22.6 billion per year. That figure is in 2018 dollars.

9. See appendix for the derivation of these figures.

10. See Pollin et al. (2023).

11. For example, Bistline et al. (2023), Penn Wharton (2023).

12. See Pollin et al. (2021) for an extended analysis of this topic.

13. That is, $3 billion per year in clean energy investments in Pennsylvania is equal to approximately 12% of $26 billion. Correspondingly, 20,000 jobs generated by $3 billion in investments is roughly 12% of 160,000 jobs generated by $26 billion in investments in the state.

14. Private health insurance coverage figures are here: https://tinyurl.com/2sewbxvx. Private sector unionization rates figures are here: https://tinyurl.com/56ua7h77

15. According to data published by the U.S. Labor Department, 20% of 65+ year-olds remain in the workforce. https://www.bls.gov/cps/cpsaat03.htm.

16. See more detailed discussions on these pension fund policies in, for example, Pollin et al. (2019).

17. https://tinyurl.com/2wf5nbxb

18. According to the 2023 article in Moneyzine "Job Relocation Expenses," these expenses for an average family range between $25,000 and $75,000 (https://tinyurl.com/yc3fhwbz). The costs include selling and buying a home, including closing costs; moving furniture and other personal belongings; and renting a temporary home or

Table A.2. Compensation Estimates for Clean Energy Sectors (2023 Dollars)

| Year | 1. Building retrofits (10,120 workers) | 2. Industrial efficiency (2,880 workers) | 3. Grid upgrades (2,610 workers) | 4. Mass transit (12,006 workers) | 5. Solar (17,640 workers) | 6. Wind (5,780 workers) | 7. Low emissions bioenergy (8,250 workers) | 8. Geothermal (6,300 workers) | 9. Small-scale hydro (7,686 workers) | Total weighted average |
|---|---|---|---|---|---|---|---|---|---|---|
| 2019 | $88,196 | $115,132 | $101,187 | $39,092 | $102,021 | $98,446 | $83,905 | $100,591 | $92,129 | $86,766 |
| 2023 | $73,100 | $100,000 | $87,700 | $39,800 | $88,400 | $85,700 | $72,300 | $84,300 | $78,100 | $75,300 |
| Average | $80,648 | $107,566 | $94,444 | $39,446 | $95,211 | $92,073 | $78,103 | $92,446 | $85,115 | $81,033 |

apartment while house-hunting for a more permanent residence. For our calculations, we assume the upper-end figure of $75,000.

19. For information about IMPLAN compensation data, see c). For information about BLS QCEW compensation data, see https://tinyurl.com/3xv539pr

## REFERENCES

Bistline, J., N. and Wolfram C. Mehrotra. 2023. *Economic Implications of the Climate Provisions of the Inflation Reduction Act.* Brookings Papers on Economic Activity. March. https://tinyurl.com/9wf5u3t6

Buchanich, Jeanine M., Evelyn O. Talbott, Vincent Arena, Todd M. Bear, James P. Fabisiak, Sally E. Wenzel, Ada O. Youk, and Jiian-Min Yuan. 2023. *Hydraulic Fracturing Epidemiology Research Studies: Asthma Outcomes.* School of Public Health, University of Pittsburgh. July. https://tinyurl.com/4m92cbzj

Chomsky, Noam, and Robert Pollin. 2020. *Climate Crisis and the Global Green New Deal: The Political Economy of Saving the Planet.* Verso.

Gorski, Irena, and Brian S. Schwartz. 2019. "Environmental Health Concerns From Unconventional Natural Gas Development." *Oxford Research Encyclopedia of Global Public Health* 25. https://doi.org/10.1093/acrefore/9780190632366.013.44

Lefebre, Ben. 2024. "Why Harris and Trump are Debating the F-word," *Politico.* September 9. https://tinyurl.com/yhpc7963

Mazzocchi, Tony. 1993. "A Superfund for Workers." *Earth Island Journal* 9 (1): 40–41.

Office of the Attorney General, Commonwealth of Pennsylvania. 2020. "Report 1 of the Forty-Third Statewide Investigating Grand Jury." https://tinyurl.com/3397zsff

Penn Wharton. 2023. "Update: Budgetary Cost of Climate and Energy Provisions in the Inflation Reduction Act." https://tinyurl.com/2z5ewkkv

Pollin, Robert. 2023. "The Political Economy of Saving the Planet." *The Japanese Political Economy* 49 (2-3): 141–168. https://doi.org/10.1080/2329194X.2023.2262531

Pollin, Robert, Jeannette Wicks-Lim, Shouvik Chakraborty, and Tyler Hansen. 2019. "A Green Growth Program for Colorado." https://tinyurl.com/mwswkuzc

Pollin, Robert, Jeannette Wicks-Lim, Shouvik Chakraborty and Gregor Semieniuk. 2021. *Impacts of the Reimagine Appalachia & Clean Energy Transition Programs for Pennsylvania: Job Creation, Economic Recovery and Long-Term Sustainability.* Political Economy Research Institute. https://tinyurl.com/mw9tmay4

Pollin, Robert, Jeannette Wicks-Lim, Shouvik Chakraborty, Gregor Semieniuk, and Chirag Lana. 2023. *Employment Impacts of New U.S. Clean Energy, Manufacturing, and Infrastructure Laws.* Political Economy Research Institute. https://tinyurl.com/y8tc4cud

# CHAPTER 5

# From Here to There—
# Advancing a Just Transition for All

TODD E. VACHON
*Rutgers University*

## Abstract

Drawing from previous research and firsthand experience in the labor movement, this chapter seeks to elaborate some of the fundamental elements of a just transition for all. For any socioeconomic transition to truly be just for workers and communities it must be protective, proactive, and transformative. *Protective* by ensuring the economic well-being of workers and communities impacted by the transition. *Proactive* by democratically developing a forward-looking transition plan through active participation by a broad base of stakeholders. *Transformative* by centering social justice, redressing past and present injustices, and recreating "the rules of the game" to ensure shared and sustainable prosperity. The key ingredients include a strong social safety net, education and training, pro-worker labor market policies, capital and revenue support, and economic democracy. Achieving any of these elements will require labor and climate activists coming out of their issue silos and connecting the dots among workers' rights, social justice, economic inequality, climate change, and democracy.

## INTRODUCTION

In 2019, General Motors (GM) confirmed the rumors that they were going to shut down their car assembly plant in Lordstown, Ohio. The plant, which opened in 1966, once employed over 10,000 workers producing combustion engine vehicles ranging from the Chevy Cavalier to GMC vans and in the end, the Chevy Cruz. As the final wave of layoffs arrived, the remaining 1,500 workers were let go, and an electric vehicle (EV) battery plant and an electric truck manufacturer were slated to open operations on the site.

In the lead-up to, and in the wake of the layoff, United Autoworkers (UAW) Local 1112 tried to help their laid-off members by setting up what they called a "just transition center" to help displaced workers search for re-employment, find training and re-education opportunities, relocate for positions in GM in other states, and navigate state and federal benefits.

Unfortunately, what it was not able to provide was what previous research has found to be the top priority of displaced workers—wage replacement and wage insurance (Parks and Baran 2023). When workers in unionized manufacturing jobs are displaced, they confront an overwhelmingly nonunion labor market with few jobs that offer equivalent pay and benefits, particularly with the educational level and skill set that was required for their previous jobs.

Economic transitions and plant closures like this have been a defining characteristic of the U.S. economy for decades, whether the result of technological changes, outsourcing, resource availability and allocation, public policy, consumer demand, or stock buybacks and other profit-seeking activities of capitalists. The subsequent mass layoffs have had serious repercussions for workers—both economically and emotionally—as well as for local economies and local tax bases (Cha, Price, Stevis, and Vachon 2021). The cycle of one unjust transition after another for U.S. workers has even weakened support for democracy itself (Leopold 2024).

While transitions like this are inherent in a capitalist economic system that's always engaging in "creative destruction," the case of Lordstown is a specific case of transition in the era of climate change and extreme inequality (Schumpeter 1950). There is no longer any doubt that climate change is negatively affecting human health and quality of life. At the same time, the top 1% of earners now take home 22% of all income in the United States, the top 10% own 76% of all wealth, real wages for U.S. workers have been stagnant for decades, and union density hovers around 10% (and just 6% for private sector workers) (Piketty 2014; U.S. Bureau of Labor Statistics 2024). The failure to act on climate and the emergence of a deregulated, oligopolistic form of capitalism are linked problems, perpetuated by the same powerful actors. We cannot successfully address one crisis without addressing the other.

This historic moment calls for bold action that simultaneously addresses both crises by putting Americans to work building an economy that is just, equitable, and sustainable (not powered by fossil fuels). The necessary transition away from an extractive economy toward a regenerative, sustainable economy will completely reshape existing labor markets, threaten to erase the gains made over generations by workers in the historically unionized blue-collar energy sector, and further shift employment into sectors where unions have been unable to gain a foothold because of a combination of employer hostility and pro-business labor laws. The fear these changes elicit in workers and communities makes solutions to the climate crisis tremendously challenging.

However, the worst possible outcomes that are feared by many—such as mass unemployment and the collapse of local economies—are only a foregone conclusion when operating under the rules of neoliberal governance. If done correctly, confronting the climate crisis offers a great opportunity to move our society toward a dual democratic commitment to work being rewarding and available to all who want it, and all workers being able to exercise voice in their workplaces and in the economy as a whole.

Achieving these ends will require a just transition plan for displaced workers as well as for historically marginalized workers and for front-line environmental justice communities. Such a plan is not just a good idea. It is an imperative—both morally and pragmatically—if we are to avoid the worst consequences of climate change and inequality. Morally, it is unjust to place the economic burden and hardship of a socially necessary transition on the workers, families, and communities that have toiled in the dirty and dangerous energy jobs that have powered the entire economy for the past century and a half. It is also unjust to further exacerbate the environmental injustices that have been forced on so many front-line communities that reaped few if any benefits from the existing carbon-based energy systems, only the harms.

Pragmatically, comprehensive climate legislation has proven to be so elusive in large part because of the inability to pull together a broad enough coalition to support it. Any group that is positioned to suffer economic losses as a result of climate mitigation is likely to be an organized force of opposition. To succeed, climate protection programs require active relationship-building among all stakeholders to develop just transition programs that address the concerns of all parties, including workers and front-line communities.

In this chapter, I draw knowledge from my previous research as well as my personal experience in the labor movement as a union carpenter, UAW organizer and local president, AFT leader, and participant in various labor–climate movement organizations to elaborate what I believe to be some of the key elements of a just transition for all. To be successful, I contend, a truly just transition for workers and communities must be protective, proactive, and transformative (Vachon 2023): *protective* by ensuring the economic well-being of workers and communities impacted by the transition, *proactive* by democratically developing a forward-looking transition plan through active participation by a broad base of stakeholders, and *transformative* by centering social justice, redressing past and present injustices, and recreating "the rules of the game" to ensure shared and sustainable prosperity.

The remainder of this chapter will briefly review the history of just transition, including its competing understandings, and outline some key elements that are needed to ensure the transition is just, moral, and pragmatically attainable—centering social justice, a strong social safety net, education and training, pro-worker labor market policies, capital and revenue supports, and economic democracy. Achieving any of these elements will require labor and climate activists to come out of their issue silos and connect the dots between workers' rights, social justice, economic inequality, climate change, and democracy.

## A JUST TRANSITION—PROTECTIVE, PROACTIVE, AND TRANSFORMATIVE

First conceived of as a "Superfund for Workers," the idea of a "just transition" originates from the work of the late U.S. labor and environmental health and safety activist Tony Mazzocchi of the Oil, Chemical, and Atomic Workers (Leopold 2007). Summarizing Mazzocchi's definition of the superfund for workers, Brecher (2015) says, "It is a basic principle of fairness that the burden of policies that are necessary for society—like protecting the environment—shouldn't be borne by a small minority, who through no fault of their own happen to be victimized by their side effects."

The term "just transition" is believed to have been first used in 1995, by Les Leopold and Brian Kohler during a presentation to the International Joint Commission on Great Lakes Water Quality (Hampton 2015). Broken into its constituent parts, "transition" refers to "the passage from one state, stage, subject, or place to another," and "just," in this usage, is the root word for justice, meaning "acting or being in conformity with what is morally upright or good." In other words, just transition combines the often-conflicting projects of economic transition and the pursuit of social justice into one combined endeavor.

When confronting the problem of climate change, the potential for injustice is great, particularly if decisions are made solely by economic elites and grounded in the logic of neoliberal capitalism (Harvey 2005; Klein 2014). This hegemonic logic of unregulated markets is at the root of the false choice that workers are confronted with when they are made to decide between good jobs or a healthy environment (Vachon and Brecher 2016). The very notion of a just transition challenges the powerful neoliberal ideology that has dominated U.S. governance since the late 1970s. It instead offers a vision of economic democracy, including public intervention into markets to account for the full social costs and benefits of environmental and economic policies to create the most just, not necessarily the most profitable, outcome for all (Cha et al. 2019). In other words, just transition puts people before profits and, by doing so, it undermines the basis for most jobs versus the environment conflicts.

Transitioning away from the current fossil fuel–based economy will require considerable economic changes, including large shifts in employment away from extractive industries and toward new green technologies and other industries. This transition will directly impact the jobs and livelihoods of tens of thousands of workers in the fossil fuel industry and indirectly affect just as many workers in related sectors such as transportation and electricity generation and transmission. New jobs created by the transition will likely require different skills and temperaments and will likely be in different geographic locations than the old fossil fuel jobs. Without intervention, the switch from fossil fuels to renewable energy will also signal a decline in traditionally unionized occupations and a shift toward jobs where unions have not yet flourished, amounting to an overall decline in bargaining power for workers in a period of already historically low levels of unionization. To illustrate this disparity in bargaining power, consider that the national average income for natural gas plant operators is $78,000 annually, while rooftop solar installers earn an average salary of just $46,000 (U.S. Bureau of Labor Statistics 2024).

At the same time, because of the legacy of racism and discriminatory hiring practices in the United States, workers from historically marginalized communities, particularly Black and Latinx workers, have often been systematically deprived of opportunities to share in the prosperity generated by the extractive economy. Adding insult to injury, these same workers have disproportionately borne the burden of the pollution created by the fossil fuel industry. For example, a recent study in the *Proceedings of the National Academy of Sciences* found that air pollution exposure in the United States is disproportionately caused by the non-Hispanic White majority but disproportionately inhaled by Black and Hispanic minorities (Tessum et al. 2019). On average, non-Hispanic White people experience a "pollution advantage" of about 17% less air pollution exposure than is caused by their consumption, while Black and Hispanic people on average bear a "pollution burden" of 56% and 63% excess exposure, respectively.

So, when asked "just transition for whom?" the answer should be "just transition for all." First, without a strong guarantee of social protections and good jobs, fossil fuel workers will be fierce opponents to legislative efforts to decarbonize the economy. Second, without good job opportunities in all industries, including healthcare, education, and services, we may find ourselves facing just as bad or worse levels of economic inequality. Third, it is not difficult to see why climate justice activists from front-line communities would be hesitant to support just transition plans narrowly focusing on helping only the well-paid fossil fuel workers when their community members were systematically denied economic opportunities while disproportionately suffering the health impacts resulting from climate destruction.

Of course, it is an oversimplification to think of these two groups as entirely separate and mutually exclusive. We know climate change harms everyone, and we know the working class is multiracial; therefore, for pragmatic reasons, climate solutions must address the concerns of those workers facing displacement. However, the patterns of inequality in access to economic opportunity and exposure to environmental burden reveal the moral need for a solution that addresses the historical legacy of racism in the United States, including how it intersects with social, environmental, and economic outcomes.

For these reasons, I have argued that a just transition program should be protective, proactive, and transformative (Vachon 2021). The protective elements of a just transition include labor market protections that provide a social floor at the workplace and community level in response to job losses in the fossil fuel and related industries as a result of decarbonizing the economy. The proactive elements of a just transition are forward-looking and involve social dialogue among various stakeholders, including labor unions and community groups, to develop a large-scale plan to transition the entire economy away from fossil fuels in the coming years. The transformative elements of a just transition involve confronting the history of injustices along the lines of race, class, and gender as well as confronting the root cause of many of these inequities— the profit motive associated with private corporate ownership and control over key elements of the economy such as natural resources and energy.

## ADVANCING A JUST TRANSITION

Overall, a just transition program must balance the needs of displaced workers organized by unions who are relatively well-off and the needs of those largely excluded from equally good employment and unions. In what follows, I briefly review what I consider to be some of the key elements of a just transition program based upon my experiences in the industry, in labor, and as a scholar. I identify six broad areas— social justice, social safety net measures, education and training, labor market policies, capital and revenue supports, and economic democracy. Most of these proposals could be actionable at the national level through legislation or executive orders; many others could be implemented by states or large municipalities or through collective bargaining agreements where unions are present. This list of potential policies provided below is not exhaustive, but for the energy transition to be just, they are necessary.

## Confront Existing Inequalities

First, a truly just transition must center social justice. To address previous and existing environmental and economic injustices, a just transition could begin by *prioritizing infrastructure projects in environmental justice communities (front-line communities affected by environmental degradation) and Indigenous communities*, including repairing lead-poisoned water supply lines, shuttering waste incinerators, closing landfills near residential neighborhoods, and replacing fossil fuel–burning power plants with renewable sources. These projects could require that a percentage of the jobs created to address these problems be filled by local residents through *local hiring requirements*.[1] Such hiring provisions could create an entry point into trade union apprenticeship programs for community members and help build careers that go beyond the immediate project at hand.

*Progressive procurement policies* could lift up women- and minority-owned businesses along the supply chain. Such businesses tend to have more diverse workforces as well, thus amplifying the social justice benefits.

To confront historical injustices in the criminal justice system, ban-the-box laws could also assist with pathways to re-entry for formerly incarcerated workers. Overall, any just transition program must redress past injustices by creating pathways for members of historically marginalized communities to participate in the new, climate-safe energy sector, whether through *pre-apprenticeship programs* in low-income community high schools and community colleges, traditional union apprenticeship programs, a federal jobs guarantee, or as a requirement for awards of public contracts to private companies bidding on projects. In other words, social justice is not just an added benefit but should be centered in the transition, including within each of the other elements reviewed below.

## Create a Strong Social Floor

Unlike most other rich capitalist democracies, the United States has a very weak social safety net. The rise of the COVID-19 pandemic in early 2020 made this abundantly clear for the world to see. Unemployment insurance payments are minimal and short in duration. The average weekly benefit for unemployed workers in 2019 was $367, and the maximum duration for receiving the benefit is 26 weeks (U.S. Department of Labor 2024). Many workers in the so-called gig economy who are classified as independent contractors are not even eligible for unemployment insurance benefits.

Further, when workers lose their jobs, they not only lose their income, but they also lose health insurance for themselves and their families if they are fortunate enough to have an employment-based private plan. Medical bills are the primary cause of housing foreclosures in the United States and lead to over 500,000 bankruptcy filings each year (Himmelstein et al. 2019). Workers also stop contributing to retirement plans when they are unemployed, further exacerbating the crisis of retirement savings among U.S. workers.

A just transition should include several improvements to and expansions of the existing social safety net to make job displacement less harmful to workers. Notably, many of these benefits would not only protect those workers displaced by climate legislation but also workers who are displaced as a result of plant closures from deindustrialization or automation—those whose jobs are devastated by the effects of climate change itself, and those who have been marginalized in the labor market as a result of structural racism, including discriminatory hiring practices and a racially biased criminal justice system.

Some social safety-net measures that could support workers in transition include *increasing unemployment insurance payment rates and extending the duration of unemployment benefits.* Going beyond unemployment insurance, a wage replacement program could provide full wages and benefits for up to three years to displaced workers as they transition to new careers.[2] Alternatively, a mandatory severance payment for workers who are part of a mass layoff could also help. Absent a more universal health program (such as Medicare for all), a stop-gap health insurance fund could be created to cover the costs of continuing displaced workers' existing employee health insurance plan for a prescribed period or until they are re-employed with a new plan, effectively decoupling healthcare from employment.

Large employers could be required to give a *minimum advance notice before a mass layoff* that would put at least 50 people out of work. Plants facing closure could also be required to *negotiate phaseout plans* with their workers. A slow and orderly phaseout plan could transition younger workers out first and into other jobs while keeping older workers employed long enough to reach their planned retirement age. Such a plan was negotiated by the electrical workers union in California during the closure of the Diablo Canyon nuclear plant. In fact, an analysis by the Political Economy Research Institute found that 83% of job losses in the fossil fuel industry can be covered through attrition by retirement (Pollin and Callaci 2019). For workers nearing retirement age, an optional *early retirement program* could be established to help those who would prefer to transition out of the workforce rather than start a new career later in life. These and other programs could be part of what Tony Mazzocchi called a "superfund for workers" to provide valuable protections for those facing job loss.

Similarly, a *universal retirement program* could be created where, when workers switch jobs, they can roll over their existing 401(k) or various other types of retirement accounts into a single universal retirement savings account, increasing overall savings by reducing paperwork burdens and financial fees for employers and employees alike. Additionally, collectively bargained *pensions should be insured* by the federal government. If banks can be bailed out, then surely workers who lose their jobs by no fault of their own can be as well.

Another idea gaining support in recent years is a *universal basic income* (UBI). The UBI, or guaranteed annual income, would be a public program providing a basic living stipend to all on an individual basis without means testing or work requirements. The income payments would be unconditional, automatic, and a right of every individual. In light of the massive unemployment experienced during the COVID-19 pandemic, support for UBI has increased, especially considering the large number of workers who were not eligible for standard unemployment insurance. The Coronavirus Aid, Relief, and Economic Security (CARES) Act provided temporary relief, including a one-time direct cash transfer along the lines of a UBI. Considering the likelihood that so many of the lost jobs during the pandemic will not return for many years, if ever, once-radical ideas like a UBI or, as proposed below, a federal jobs guarantee, are worth serious consideration. However, to be just, a UBI would have to be part of a larger package of social welfare programs and labor market policies— not a substitute for them—because cash transfers without structural changes do not help solve problems in the long run.

## Retrain and Educate

In the United States, education is an expensive endeavor that places nearly all the risk, usually in the form of debt, on the individual, with no guarantee that the investment will lead to a job in the end. The transition for workers from one occupation to another is made easier when education is highly subsidized or free and, like health insurance and other benefits, continues during interim periods of unemployment. The lack of affordable education options and adequate social welfare programs has been cited as one of the major reasons for the difference in support for climate protection measures between workers in the European context and the U.S. context (Hyde and Vachon 2018). The following are some education measures that could make socioeconomic transitions less harmful to workers.

First and foremost, a just transition would highly *subsidize job training*, retraining, and college education—including at technical schools and liberal arts colleges—for displaced workers and workers from historically marginalized communities. Such efforts at decommodifying education could also go a long way toward reducing the historic disparity in educational opportunities for children from affluent versus less affluent families. Additionally, the opportunity for individuals to access higher education and advanced training will support the development of specific skills needed for building, operating, maintaining, and managing the new green economy.

Existing *apprenticeship programs* could be expanded as well, and new ones created for occupations that are crucial to the ongoing energy transition and the future maintenance of a regenerative economy. From wind to solar to storage technologies to EV charging, mass transit, and smart grids, the investment in training will not only help to transition displaced workers but will also be necessary to successfully build and operate a sustainable economy. Apprenticeships can also create pathways into unionized jobs for workers who have been historically excluded from such opportunities. Training and education programs should also be created to help expand already-existing low-carbon sectors of the labor market, such as healthcare and education. Further, these and other programs should specifically *target front-line communities* that have systematically encountered a dearth of good education and employment opportunities.

## Empower Workers

One of the key fears of many blue-collar workers is that the collapse of the fossil fuel industry will also mean the collapse of many good, unionized job opportunities for workers with less than a college degree. Globalization and outsourcing have already stripped away most of the good-paying manufacturing jobs that were a pathway to the middle class in a previous era. The fossil fuel industry, electric utilities, and commercial construction are some of the few remaining areas where unions still thrive in the private sector. Another concern is the geographic specificity of fossil fuel work and the likelihood that new jobs would be in different locations, requiring different skills. A strong package of labor market protections and regulations could help unions to grow again, to make jobs in all industries into good jobs, and to help create jobs locally in the regions where old jobs are lost. Some parts of such a package include living-wage ordinances, changes to union election rules, expansion of prevailing-wage laws, local hiring and procurement measures, and a federal jobs program where the government could serve as the employer of last resort by hiring unemployed workers to engage in climate protection work or to address the many social needs not currently met by the private sector. More details on each are below.

Replacing the minimum wage with *a living-wage law* that accounts for variation in the cost of living across space and time would go a long way to ensure that workers transitioning away from fossil fuels are not left behind. A living-wage provision would also ensure that existing low-wage jobs, which are disproportionately performed by women and people of color, pay a fair rate. Many of these jobs—often in the service and care sectors—are also key elements of a carbon-neutral economy, but they have been devalued in part because of the gender composition of the workforce combined with the history of pay inequity for women and people of color. In addition to raising wages, *reducing working hours* would help to ensure adequate employment opportunities for all who want to work.

*Prevailing-wage laws,* such as the Davis–Bacon Act, ensure that workers employed on public construction projects receive family-supporting compensation reflective of the local market rate. The main purpose of prevailing-wage laws is to maintain local labor quality, training, and wage standards and to support the local economy in the competitive public bidding process. The law creates a level playing field for all contractors by ensuring that workers are paid an appropriate wage based on job classification and skill, incentivizing contractors to compete over factors other than labor costs. A *green prevailing-wage* law could be created to set locally determined minimum wage and benefit requirements for all green transition–related employment, ranging from construction to operations and services. A green prevailing wage could also be used to set sector-level wages to ensure that renewable energy jobs pay the same rate as fossil fuel jobs.

Research has attributed much of the growth of income inequality in recent decades to the decline in unionization (Hyde, Vachon and Wallace 2017; Western and Rosenfeld 2011). *Card-check union recognition* would make unionization less burdensome and undermine many of the anti-union tactics that employers currently deploy with little or no penalty. Currently in the private sector, workers must petition to hold an election and then wait several weeks before the election is held, a period in which employers engage in intimidation campaigns to drum up fear and break unionization efforts. Card-check recognition would allow unions to be formed when a majority of members sign cards indicating they would like to form a union.

A *first contract statute* could also be created to ensure companies cannot break newly formed unions by denying a first contract. Under such a provision, employers would be required to begin negotiating in good faith within a defined number of days of receiving a request from a newly formed union. If no agreement is reached after a defined period of negotiation, the parties would begin a compulsory mediation process. If an agreement is not reached after the mediation period, the contract would be settled through binding arbitration.

Additional provisions to help unions grow include a partial or complete *repeal of the Taft–Hartley Act*. For example, Section 14(b) of the act has allowed 28 states to pass so-called right-to-work legislation that eliminates the ability of unions to collect dues from those who benefit from union contracts and activities, creating the potential for large-scale free-rider problems. Taft–Hartley also outlawed jurisdictional strikes, wildcat strikes, solidarity strikes, political strikes, secondary boycotts, and mass picketing, all of which undermine the power of an organized labor force to fight for and win major gains for the working class.

A just transition should also extend the right to organize to all categories of workers. Currently, many workers, such as domestic workers and farmworkers, lack the basic protections that other workers enjoy under federal labor law.[3] A just transition could also ensure that all employment relationships are properly classified as such. Many of these jobs will form a solid base of the low-carbon economy, and workers in these sectors should be allotted the same rights as workers in other sectors. Most of the so-called gig economy misclassifies employees as independent contractors to avoid providing benefits and undermine the ability of workers to unionize. An A-B-C test of employee status could help ensure that workers are properly classified as employees for the purposes of collective bargaining.[4] The Protecting the Right to Organize Act (PRO Act), passed by the U.S. House of Representatives in early 2021, would go a long way to address all the labor law issues raised here and strengthen the ability of workers to organize and bargain collectively.

When it comes to facility closures as a result of decarbonization, or for any other reason, employers should be required to *bargain over impacts* with workers whether they have a union in place or not. Workers should also have the right to purchase their workplaces before any other buyers when they are up for sale or slated to close—a process known as the *right of first refusal*. Proponents of worker ownership have proposed that companies being sold or closed should be held in escrow for a period while the workers are made aware of how much they would need to pay to exercise the right of first refusal (Gowan 2019). The workers would then vote on whether to pursue the purchase. Special funds, described more in the next section, could be created to help facilitate these transitions of ownership.

*Local hiring statutes* could mandate that green transition projects hire a certain percentage of workers from the community in which the project is being completed. This would help to create pathways into jobs with career ladders for historically marginalized groups of workers. As these workers gain skills and experience, they will earn living wages doing the work of redressing a long history of environmental racism in their local communities. *Local procurement statutes* would require a certain percentage of materials used for green transition projects to be produced locally. This would help to revitalize domestic manufacturing and create additional blue-collar job opportunities for workers with less than a college degree (BlueGreen Alliance 2019).

Perhaps the boldest active labor market policy would be *a federal jobs guarantee.* A "jobs for all" plan could be a federal program similar to the original New Deal's Works Progress Administration that provides funds for local governments, nonprofit organizations, and other agencies serving the public to employ anyone who wants a job at a living wage. This type of Keynesian full employment policy would not only ensure jobs for all who want them but would likely be necessary to mobilize the productive capacity needed to build a truly climate-safe society. The jobs program would not exclude any individual or group of people who want to work, and it could be transformative by making special efforts to employ workers from front-line communities, veterans, at-risk youth, ex-convicts, people with disabilities, and other people with special needs and/or barriers to employment.

## Support Communities

Job reductions can affect more than just individual workers. Often, whole communities suffer from the decline in the spending by displaced workers as well as the decline in tax revenue at the local level. The problem is particularly germane in many fossil fuel extraction communities with less diversified economies that rely heavily on the income and tax revenue generated by a small number of employers in just one industry—such as coal mining. In these cases, a *tax base support fund* could provide money to the municipality or county to replace lost tax revenues and keep vital services, such as the public school system, operating for up to five years while a broader economic transition plan is developed (Lipsitz and Newberry 2017). Such a plan could draw lessons from the highly successful process that helped local communities adjust to the disruption caused by military base closures under the 2005 Base Realignment and Closing Commission (Brecher 2015). Those communities were supported by a variety of federal assistance programs, including planning and economic assistance, environmental cleanup, and funding for a range of community services.

*Community development grants* are another effective means of funding efforts by communities to reshape and diversify their local economies and transition to a post-carbon income and tax base.

*Low- or zero-interest loans* for new businesses can also help to jump-start new economic ventures in former fossil fuel communities. Particular focus should be given to cooperatives and worker-owned enterprises, which tend to lack a good source of start-up capital. In the case of workers seeking to purchase their place of employment under the "right of first refusal" protection described above, they should be able to access a *dedicated pool of capital for worker ownership transitions.* The loan could be repaid with revenue generated by the now cooperatively owned enterprise or via payment in kind through some combination of efforts by the cooperative to address social needs through production, being more environmentally sustainable, or rectifying legacies of discrimination in the workplace and/or community (Gowan 2019).

## Expand Democracy

When faced with the option of either keeping well-paying fossil fuel jobs or putting faith into hypothetical plans for a transition, it's not hard to understand why many workers have fought to protect their jobs—especially if they have not been included in discussions around what a transition for them would look like (Isser 2020). *Stakeholder participation in designing a just transition,* including democratic and participatory processes that are inclusive of and led by workers and front-line communities, are paramount to the initial political success and ultimate economic success of any transition plan. One way to ensure the process is democratic and inclusive would be for all participating parties to *adopt the Jemez principles for democratic organizing* as ground rules for the difficult discussions that will be required to develop a just transition program (see Southwest Network for Environmental and Economic Justice 1996).

A just transition should also help to reduce the broader democracy deficit that workers and front-line communities face in the United States and which often leads to the false choice between jobs and environmental protections. Mandating *worker membership on corporate boards* is one way to ensure that workers have some influence in their companies' decision-making processes. Broadening and deepening the scope of collective bargaining is another way of empowering workers on the job and in the economy. For example, *sectoral bargaining* would allow workers to bargain wages and working conditions across an entire sector involving multiple employers. The 1938 Fair Labor Standards Act originally allowed for sectoral bargaining in some industries, but that aspect of the law was repealed in the late 1940s (Andrias 2019). Sectoral bargaining could be included in just transition plans to ensure that new green energy jobs offer the same rates and benefits as the old fossil fuel jobs.

Bargaining for the common good (BCG) seeks to expand democracy by reimagining the participants, processes, and purposes of collective bargaining (McCartin, Smiley, and Sneiderman 2021). BCG efforts *expand the scope of bargaining* to include demands that go beyond wages and benefits and *expand the participants at the bargaining table* by bringing union members and community allies together to develop demands jointly. The union and its allies then present their demands not only to the direct employer of the unionized workers but also to the web of corporate and financial relationships that influence or control the employer. Most importantly, BCG efforts see the purpose of collective bargaining as not only securing material gains for workers but also addressing deeper structural inequities and building a more just and sustainable world. Changes in existing labor law could expand the scope of mandatory topics for bargaining and/or expand the bounds of bargaining unit membership to include vital stakeholders in the process.

The existing, neoliberal approaches to the climate crisis have repeatedly undermined democracy by putting private profits before the common good. The phrase *energy democracy* has emerged in recent years as part of a transformative vision of just transition. The core tenet of energy democracy is the transfer of ownership and control of the energy sector to the public, which could be local communities, or to new and/or "reclaimed" public entities such as utilities (Vachon and Sweeney 2018). Proponents such as Trade Unions for Energy Democracy argue that the only way to ensure a just transition to renewable energy sources is to replace the profit motive as the primary factor in decision making regarding energy and replace it with the public interest. Publicly owned and democratically controlled energy systems whose guiding principle is maximizing the public good will facilitate the rapid transfer to renewable sources that is needed to address climate change and can ensure that the transition is done equitably, protecting vulnerable workers and addressing existing environmental injustices.

## TOWARD A DEMOCRATIC AND SUSTAINABLE FUTURE

A few years after the Lordstown auto plant closed in Ohio, the Ultium EV battery plant opened, employing approximately 1,800 workers. When production began in 2022, the starting pay was $16 per hour, just a fraction of the starting pay at the former GM plant which was $27 per hour three years earlier. Workers in the new plant quickly organized with the UAW and were able to negotiate some wage and benefit increases at the local level, but the nationwide "stand-up" auto strike in the summer of 2023 made major advances for the EV plant workers. Through organizing and demanding a just transition that folded battery plant workers into the GM master agreement, the UAW was able to raise starting salaries at Ultium to the same as the autoworkers who were previously employed at Lordstown—but for just a fraction of the number of workers and after three years without good jobs.

To be truly just for all, I argue, the energy transition must be protective, proactive, and transformative. For both moral and pragmatic reasons, it must prioritize social justice and the protection of displaced workers, and it must be developed democratically through active participation by a broad base of stakeholders. Through a federal jobs guarantee, free or subsidized education, and various innovations in labor market policy that focus on social justice, all people in the United States who want to work could be gainfully employed and earn a living wage with ample benefits. Low-carbon workers such as nannies, housekeepers, and caregivers would be valued and paid adequately for their important work in a regenerative economy. Displaced fossil fuel workers would be re-employed at equivalent wages, and historically marginalized populations would have equal access to new and existing labor market opportunities.

With full employment in effect, the dirtiest and most dangerous jobs would have to be improved to attract employees, or else technology would be developed to perform those tasks. The public jobs program would also prioritize projects that are deemed socially valuable rather than those that are most profitable, leading to a net improvement in the quality of life for everyone.

Improvements in the social safety net—including the decoupling of healthcare from employment—and changes to corporate governance structures and union bargaining laws would dramatically increase economic democracy. Workers would have greater freedom to pursue the careers they are passionate about without worrying about the health benefits package offered by various employers. They would also be free to switch jobs throughout their lives without fear of losing vital benefits. Through their seats on corporate boards, workers would exercise greater control over economic decisions made at the firm level. Through deepened and broadened collective bargaining, workers would also enjoy greater influence over decisions affecting the broader economy.

Overall, a socially just transition program could be imagined as an updated and expanded version of the GI Bill of Rights that provided education and training loan as well as guarantees for homes, farms, and businesses, in addition to unemployment pay for veterans returning from World War II. A similar Green Bill of Rights is needed today for members of front-line communities and those who are displaced from their jobs through no fault of their own. Just like the broader New Deal analogy, the bill-of-rights program would have to be updated not only to include but to prioritize the needs of folks who have been historically marginalized throughout the 19th and 20th centuries, including people of color, Indigenous peoples, women, the disabled, and other groups that have often been left behind in government social programs.

In this chapter, I have laid out what I see as six key elements of a truly just transition that would need to be included in such a bill of rights—centering social justice, a strong social safety net, education and training, pro-worker labor market policies, capital and revenue supports, and economic democracy. A just transition program that incorporates these six elements, thereby challenging the dominant free market ideology and expanding and deepening democracy in the United States, is both morally and pragmatically necessary to successfully confront the dual crises of climate change and inequality. Only by putting people before profits and well-being before bottom lines can we fundamentally reorient our economy and move toward sustainable and shared prosperity. This will require massive bottom-up organizing, coming out of our issue silos, and connecting the dots among workers' rights, social justice, economic inequality, climate change, and democracy. It will also require political and electoral work at all levels of government. It is not a simple task, but as Frederick Douglass reminds us, "If there is no struggle, there is no progress. ... Power concedes nothing without a demand. It never has and it never will" (Douglass 1857).

## ENDNOTES

1. This requires overturning a Reagan-era ban on geographically based hiring preferences for federally funded projects.

2. Pollin and Callaci (2019) estimate that a program for all workers and communities that are currently dependent on domestic fossil fuel production would cost $600 million per year to pay for: (1) income, retraining, and relocation support for workers facing retrenchments; (2) guaranteeing the pensions for workers in the affected industries: and (3) effective transition programs for fossil fuel–dependent communities.

3. These categories of predominantly Black workers were excluded at the time from the National Labor Relations Act in 1935 to garner the required votes from southern Democrats in Congress.

4. Some courts using this test look at whether a worker meets three separate criteria to be considered an independent contractor: (1) the worker is free from employer's control or direction in performing the work, (2) the work takes place outside the usual course of the business of the company and off the site of the business, and (3) the worker is engaged in an independent trade, occupation, profession, or business.

## REFERENCES

Andrias, Kate. 2019. "A Seat at the Table: Sectoral Bargaining for the Common Good." *Dissent*. March. https://tinyurl.com/yuvkptxr

BlueGreen Alliance. 2019. "Solidarity for Climate Action." https://tinyurl.com/k2437yw5

Brecher, Jeremy. 2015. "A Superfund for Workers: How to Promote a Just Transition and Break Out of the Jobs vs. Environment Trap." *Dollars and Sense*. November/December. https://tinyurl.com/3a7um8ux

Cha, J. Mijin, Manuel Pastor, Madeline Wander, James Sadd, and Rachel Morello-Frosch. 2019. "A Roadmap to an Equitable Low-Carbon Future: Four Pillars for a Just Transition," The Climate Equity Network. April. https://tinyurl.com/ynvmyebv

Cha, J. Mijin, Vivian Price, Dimitris Stevis, and Todd E. Vachon. 2021. "Workers and Communities in Transition: Report of the Just Transition Listening Project." Labor Network for Sustainability. https://tinyurl.com/3me5zb4b

Douglass, Frederick. 1857. "Frederick Douglass Declares There Is 'No Progress Without Struggle.'" SHEC: Resources for Teachers. https://shec.ashp.cuny.edu/items/show/1245

Gowan, Peter. 2019. "Workers Should Be in Charge." *Jacobin*. April 17. https://tinyurl.com/2vmjbv7z

Hampton, Paul. 2015. *Workers and Trade Unions for Climate Solidarity: Tackling Climate Change in a Neoliberal World*. Routledge.

Harvey, David. 2005. *A Brief History of Neoliberalism*. Oxford University Press.

Himmelstein, David U., Robert M. Lawless, Deborah Thorne, Pamela Foohey, and Steffie Woolhandler. 2019. "Medical Bankruptcy: Still Common Despite the Affordable Care Act." *American Journal of Public Health* 109, 431–433. https://tinyurl.com/3w6n4r9p

Hyde, Allen, and Todd E. Vachon. 2018. "Running With or Against the Treadmill? Labor Unions, Institutional Contexts, and Greenhouse Gas Emissions in a Comparative Perspective." *Environmental Sociology* 5, no. 3, 269–282. https://doi.org/10.1080/23251042.2018.1544107

Hyde, Allen, Todd E. Vachon, and Michael Wallace. 2017. "Financialization, Income Inequality, and Redistribution in 18 Affluent Democracies." *Social Currents* 5, no. 2, 193–211. https://doi.org/10.1177/23294965177048

Isser, Mindy. 2020. "Joe Biden Thinks Coal Miners Should Learn to Code. A Real Just Transition Demands Far More." *In These Times.* January 15. https://tinyurl.com/yy9nc4bm

Klein, Naomi. 2014. *This Changes Everything: Capitalism vs. the Climate.* Simon and Schuster.

Leopold, Les. 2007. *The Man Who Hated Work and Loved Labor.* Chelsea Green.

Leopold, Les. 2024. *Wall Street's War on Workers: How Mass Layoffs and Greed are Destroying the Working Class and What to Do About It.* Chelsea Green.

Lipsitz, Richard, and Rebecca Newberry. 2017. "The Huntley Experiment: Building Strategic Alliances for Real Change." Labor Network for Sustainability. https://tinyurl.com/fdjsk727

McCartin, Joseph A., Erica Smiley, and Marilyn Sneiderman. 2021. "Both Broadening and Deepening: Toward Sectoral Bargaining for the Common Good." In *Revaluing Work(ers): Toward a Democratic and Sustainable Future,* edited by Tobias Schulze-Cleven and Todd E. Vachon. Cornell University Press.

Parks, Virginia, and Ian Baran. 2023. "Fossil Fuel Layoff: The Economic and Employment Effects of a Refinery Closure on Workers in the Bay Area." University of California, Berkley Labor Center. April. https://tinyurl.com/552d6yk9

Piketty, Thomas. 2014. *Capital in the 21st Century.* Harvard University Press.

Pollin, Robert, and Brian Callaci. 2019. "The Economics of Just Transition: A Framework for Supporting Fossil Fuel-Dependent Workers and Communities in the United States." Labor Studies Journal 44, no. 2. https://doi.org/10.1177/0160449X18787051

Schumpeter, Joseph A. 1950. "The Process of Creative Destruction." In *Capitalism, Socialism and Democracy,* 3rd ed., edited by Joseph A. Schumpeter. Allen and Unwin.

Southwest Network for Environmental and Economic Justice. 1996. Jemez Principles for Democratic Organizing. December. https://www.ejnet.org/ej/jemez.pdf

Tessum, Christopher W., Joshua S. Apte, Andrew L. Goodkind, Kimberley A. Mullins, David A. Paolella, Stephen Polasky, Nathaniel P. Springer, Julian D. Marshall, and Jason D. Hill. 2019. "Inequity in Consumption of Goods and Services Adds to Racial–Ethnic Disparities in Air Pollution Exposure." *Proceedings of the National Academy of Sciences of the United States of America* 116, no. 13, 6001–6006. https://doi.org/10.1073/pnas.1818859116

U.S. Bureau of Labor Statistics. 2024. "Occupational Employment Statistics." https://tinyurl.com/zarjeysb

U.S. Department of Labor. 2024. "About Unemployment Insurance." https://tinyurl.com/392f24jr

Vachon, Todd E. 2021. "The Green New Deal and Just Transition Frames within the American Labor Movement." In *Handbook of Environmental-Labour Studies,* edited by Nora Räthzel, Dimitris Stevis, and David Uzzell. Palgrave Macmillan.

Vachon, Todd E. 2023. *Clean Air and Good Jobs: U.S. Labor and the Struggle for Climate Justice.* Temple University Press.

Vachon, Todd E., and Jeremy Brecher. 2016. "Are Union Members More or Less Likely to Be Environmentalists? Some Evidence from Surveys." *Labor Studies Journal* 41, no. 2: 185–203. https://doi.org/10.1177/0160449X16643323

Vachon, Todd E., and Sean Sweeney. 2018. "Energy Democracy: A Just Transition for Social, Economic, and Climate Justice." In *Agenda for Social Justice: Global Solution,* edited by Glenn W. Muschert, Brian V. Klocke, Robert Perucci, and Jon Shefner. Policy Press.

Western, Bruce, and Jake Rosenfeld. 2011. "Unions, Norms, and the Rise in US Wage Inequality." *American Sociological Review* 76, no. 4, 513–537. https://doi.org/10.1177/0003122411414817

# Climate Jobs and Manufacturing: Green Industrial Policy Must Mean Good Jobs

MIKE WILLIAMS

*Center for American Progress*

## Abstract

The movement toward a climate-conscious economy in the United States must incorporate two key elements: labor standards attached to climate policies and investments linked to industrial policies that build out domestic clean-tech supply chains. Both efforts support a strong middle class, and both shined through in the Inflation Reduction Act. High-quality union construction jobs have seen a significant uptick in clean energy industries that had long bristled at using union labor, and investment announcements for new manufacturing facilities and capacity seemed to crop up nearly every day over the past year—until the Trump administration began to eviscerate the Biden-era climate and energy policies. There is, however, another problem within the good news of jobs and clean energy. The improvement in the quality of jobs—notably the expansion of union density—that the construction sector has seen in clean-tech industries has not had nearly the same success in the manufacturing sector. Industrial policy is incomplete without high-quality labor standards for production, operations, and maintenance workers.

## SCENE SETTING WITH BACKROOM DEALING THAT SCREWS WORKING PEOPLE

As the end of July 2022 approached, a breakthrough occurred. Senators Joe Manchin and Chuck Schumer emerged with a framework deal (Romm, Stein, Roubein, and Joselow 2022) that would become the largest and among the most consequential measures of the Biden presidency and the most impactful piece of climate policy—possibly in history up to this point (Energy Innovation and Marcacci 2022) (hopefully to be overtaken in that ranking sometime, somewhere soon). The story here is not about what they agreed on—the Inflation Reduction Act (IRA). That is a well-trodden path; the story is about what was traded away and why the United States still has a huge hole when it comes to job quality in manufacturing.

Joe Manchin is from West Virginia. Toyota has a combined engine and transmission plant in his state (Toyota Newsroom, no date) and is staunchly opposed to its workers unionizing. Passing the legislation that eventually turned into the IRA required support from all 50 Democratic senators—because not one Republican would engage in good faith. Each senator had the ability to hold up the bill for whatever they wanted, and a few of them took advantage of this leverage—Manchin perhaps the most. He had many demands throughout the months of negotiations, and given his leverage, all his demands were met.

One of the earlier demands was to eliminate, or at least strip down, the tax credit for electric vehicles (EVs). What was not fully known was that Manchin was specifically targeting an additional benefit for EVs built with union labor. In the proposed legislation, then known as Build Back Better, an additional $4,500 would have been stacked on top of the rest of the credit if the workers at a facility that assembled the EV (within the United States) worked under a collective bargaining agreement (CBA) (Build Back Better Act 2021). This provision would have been a direct (and hefty) incentive on behalf of quality union jobs in domestic manufacturing.

It also specifically aimed to strengthen the hand of U.S. autoworkers and change the downward trend of union density within the domestic auto industry—a trend that has only been strengthened by companies like Ford and General Motors taking advantage of decades of pro-business, neoliberal trade policies and foreign-owned companies arriving on our shores with mostly hostile attitudes toward worker organizing. The biggest new entrant to the EV industry, Tesla, has also taken up that hostility, with billionaire owner Elon Musk famously offering up frozen yogurt instead of collective bargaining rights to Tesla's workers (O'Donovan 2021).

The whole credit was not contingent on a CBA, just the addition. The point was to incentivize EVs and make the incentive more potent if the car was built by workers with a good job and a fair wage. This was apparently too much for Manchin (and Toyota and Musk) (Milman 2021). In Manchin's comments and complaints to the press and industry groups, the credit was too rich for the federal government (Davenport, Friedman and Tabuchi 2022), but interestingly, he never openly talked about the union bonus. Behind closed doors though, the axe fell for what would have been the first targeted incentive based on work done under a CBA within the United States federal tax code.

And with that, the only congressionally mandated hook for incentivizing good union jobs within manufacturing disappeared from President Biden's landmark legislation. It is an example of the gap we face in ensuring that manufacturing jobs are high-quality union jobs, and most especially as we undertake the massive transition to a climate-conscious economy.

## THE JENGA-IZATION OF THE ECONOMY

Manufacturing supports local communities and often provides quality, middle-class livelihoods for working people. According to the Economic Policy Institute,

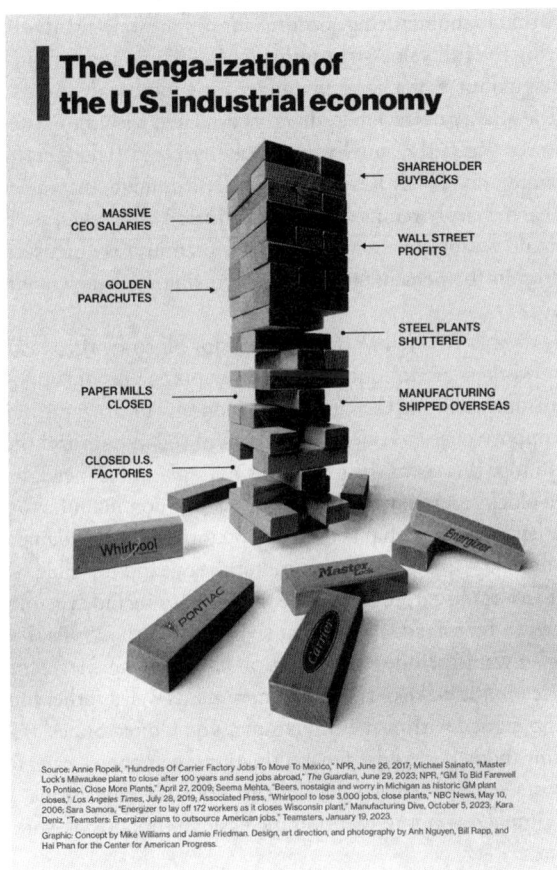

**The Jenga-ization of the U.S. industrial economy**

SHAREHOLDER BUYBACKS

MASSIVE CEO SALARIES

WALL STREET PROFITS

GOLDEN PARACHUTES

STEEL PLANTS SHUTTERED

PAPER MILLS CLOSED

MANUFACTURING SHIPPED OVERSEAS

CLOSED U.S. FACTORIES

Whirlpool   Energizer   Master   PONTIAC   Carrier

Source: Annie Ropeik, "Hundreds Of Carrier Factory Jobs To Move To Mexico," NPR, June 26, 2017; Michael Sainato, "Master Lock's Milwaukee plant to close after 100 years and send jobs abroad," The Guardian, June 29, 2023; NPR, "GM To Bid Farewell To Pontiac, Close More Plants," April 27, 2009; Seema Mehta, "Beers, nostalgia and worry in Michigan as historic GM plant closes," Los Angeles Times, July 28, 2019; Associated Press, "Whirlpool to lose 3,000 jobs, close plants," NBC News, May 10, 2006; Sara Samora, "Energizer to lay off 172 workers as it closes Wisconsin plant," Manufacturing Dive, October 5, 2023; Kara Deniz, "Teamsters: Energizer plans to outsource American jobs," Teamsters, January 19, 2023.

Graphic: Concept by Mike Williams and Jamie Friedman. Design, art direction, and photography by Anh Nguyen, Bill Rapp, and Hai Phan for the Center for American Progress.

manufacturing workers—who make up more than 11 million people in the U.S. workforce—earn 13% more in hourly compensation than comparable workers in other industries, and they have an advantage in health care and retirement benefits (Mishel 2018).

Manufacturing's impacts on the broader economy are foundational and yet often understated (Gold 2016). The act of producing a good has a long stream of value, from processing the raw materials through production and on into the downstream sales. An analysis that considers the value from inputs shows that manufacturing accounts for more than 11% of the U.S. gross domestic product (GDP), including a total output of more than $2.3 billion in 2018 (National Association of Manufacturers 2019). Research shows that these numbers may be lower than the reality, as they underestimate the "multiplier effect" that manufacturing jobs bring to local communities, as well as the longer-term benefits that accrue to the families of appropriately paid manufacturing workers (MAPI Foundation 2016). The MAPI research shows that manufacturing accounts for roughly one-third of the U.S. GDP when considering the full value-stream impact (Williams and Sutton 2021).

And yet the guarantee of a good manufacturing job has dwindled over the past few decades. The quality of the job is directly correlated with the existence of unions and worker power. Recently stated very clearly by Dean Baker, co-founder of the Center for Economic and Policy Research, "There is little doubt that the manufacturing premium has been sharply reduced, if not eliminated altogether, over the last four decades. The main reason for the decline in the premium is not a secret. There has been a huge drop in the percentage of manufacturing workers who are unionized" (Baker 2024).

Why is this the case? A well-known game helps illustrate the reasons why fewer manufacturing workers are unionized. The overlapping eras of supposed free trade and anti-unionism turned the U.S. industrial economy into a game of Jenga. Blocks of wealth and opportunity have been ripped out of the middle and the bottom and then stacked on top. But, as in the game, it is not only unbalanced, it all inevitably falls apart. The blocks representing the loss of good union manufacturing jobs hold the tower together, and those that get stacked on top—CEO golden parachutes and shareholder buybacks—make the structure top-heavy and unstable.

The era of free trade can roughly be considered to include the implementation of North American Free Trade Agreement (NAFTA) in the 1990s through China's accession to the World Trade Organization in the early 21st century and the reverberations felt in the wake of those major events, as well as other problematic free trade agreements signed within the time frame. The United States' trade policy left its manufacturing industries exposed to multinational companies eager to find cheaper production costs at the expense of decent working conditions and negative environmental impacts—not to mention the basic livelihoods of working people in the United States.

The *New York Times*'s "The Daily" podcast ran an episode on the impact of NAFTA on U.S. politics and told the story of Chancie Adams, a disaffected man who once had a promising career at Master Lock in Milwaukee (Barbaro and Kaufman 2024). That career was taken away from him and so many others because the company decided to move production to Mexico, first shipping over 1,000 union jobs out of the country immediately after NAFTA passed and then slowly grinding down its domestic operations until the final pieces of production were shut down in early 2024. Joe Wrona's experience in Buffalo is not so different (Wrona 2021). He spent years working at a facility that made metal silicate, which happens to be a base component of polysilicon—core to the production of photovoltaic solar panels. His company made investments and significant plans to integrate into this global supply chain only to be completely undercut by competition from China that was relying on forced labor to produce polysilicon and solar panels.

Preceding and overlapping the era of supposed free trade has been the concerted effort to break unions and discourage the efforts of working people to organize across the country. In fact, it was the ease with which companies could move production to ever-cheaper locations that afforded them the opportunity to stand in the way of unionization and better benefits for their employees.

As a reminder of why unions and collective bargaining are important, Lawrence Mishel from the Economic Policy Institute writes,

> Collective bargaining increases and equalizes wages for union workers and nonunion workers in unionized occupations and sectors. Wage data has long demonstrated the connection between being represented by a union and earning higher wages. This advantage, called the "union wage premium," measures the percent difference between the wages of unionized workers and those of non-unionized workers with the same characteristics. (Mishel 2021)

The busting and avoiding of unions have a complementary relation to the movement of jobs overseas. So-called right-to-work laws—birthed in the Jim Crow South to break multiracial organizing—spread throughout the region and have been a stalwart demand of anti-union advocates. It just so happens that when manufacturing CEOs decided that they wanted to find cheaper labor, they had another alternative to shipping jobs out of the country. The impact of these "right-to-work" laws created fertile ground for low-wage jobs and weak unions in the U.S. South. Many manufacturers located in northern states took note and immediately began a decades-long effort to relocate where they could employ lower-wage (and less-protected) workers.

## MODERN INDUSTRIAL POLICY

The role of government to act directly within and as an influence on the economy has been debated in politics for centuries. Industrial policy—a critical tool of macroeconomic policy—has been deployed to build up core industries important to national identity as well as the local economy, to lift nations out of the depths of depressions and recessions, and to rebuild from the ashes of war. It is a hallmark of developing nations striving to achieve middle-class livelihoods for their people. Plainly speaking, industrial policy is when a government supports an industry, whether through trade policy, direct incentives or investment, and/or research and development.

Examples abound throughout history, from the debates between Alexander Hamilton and Thomas Jefferson on the creation of a national bank that would support domestic manufacturing (Hill 2015)) to the rebuilding efforts of the United Kingdom and France after World War II that aimed to solidify and expand core industries such as steel and automobiles and to develop new industries such as computing (Owen 2012). In the past few decades of the 20th century, Japan, South Korea, and other nations leaned heavily into building out capacity for advanced technology manufacturing (Sarel 1996), while China launched history's most ambitious industrial policy at the beginning of the 21st century—taking advantage of lax trade policy—which has turned it into a global manufacturing behemoth (Jigang 2020).

At the beginning of the 21st century, industrial policy took a back seat to market-based tools such as trade liberalization in industrialized economies. The recession of 2008 brought it back momentarily, notably as we saw with the U.S. government's intervention to prop up the financial sector and save the domestic auto industry. It soon lost traction in economic debate, but now we sit at a time where—possibly owing to COVID-19's impacts on national and global economies—the method and breadth of industrial policy, not its use or existence, is the debate. This is summed up well by Todd Tucker of the Roosevelt Institute when he says,

> After decades of policymakers attempting to minimize the actual or perceived role of the state in the economy, the state is undeniably back as a key actor. State-dominated countries like Russia and China make daily headlines. States are waging wars against one another and against non-state actors around the globe. Closer to home, Presidents Trump and Biden have used emergency powers from the FDR era to tackle everything from manufacturing vaccines to the baby formula shortage to deploying heat pumps. The US has also deployed tax credits, loans, and grants on a scale not seen in generations. (Tucker 2024)

This is a welcome shift in the economic debate—especially as the uptake of industrial policy is directly correlated with solutions to the climate crisis in many nations around the world. The IRA in the United States is perhaps the most potent example, but others have followed suit. The European Union moved forward with its Green Deal Industrial Plan, a €100 billion effort focused on streamlining regulatory processes, building up export markets and shoring up critical mineral supply chains, and directly supporting clean technology industries (European Commission, no date). Japan is in the midst of deploying its Green Transformation, a 150 trillion yen investment (public and private) that aims to change Japan's energy sector while also spurring the rest of Southeast Asia to change alongside it (and securing key export markets) (Cabinet Secretariat of Japan 2023). And Brazil embraced the work of Italian economist Mariana Mazzucato—noted for her work on building out 21st-century industrial policy—by establishing six missions for Brazil's industrial strategy and dedicating 300 billion Brazilian real, which includes a dedication to improving and building out public transport and decarbonization of heavy industry within its broader energy transition (Brazil Presidency 2024).

Brazil's industrial policy strategy—Nova Industrial Brasil—explicitly calls for the promotion of better jobs. The EU's Green Industrial Plan includes efforts on reskilling and upskilling. Efforts from other nations barely mention jobs and working people and often completely ignore the quality of jobs created by industrial policy. The IRA in the United States put a serious focus on the quality of jobs in the construction sector—but nothing for manufacturing work. This is an explicit hole in green industrial policy, in spite of warnings and suggestions from academics and official institutions. The United Nations International Labour Organization published a report with the following clear recommendation:

> Industrial policies need to be designed with a view to fostering structural transformation patterns that have the potential to accelerate the generation of not just more jobs, but also more productive and better jobs. (Salazar, Nübler, Kozul-Wright, and International Labour Office 2014).

Over the past few years, the United States started to build a more solid foundation with better trade enforcement and the beginning of a concerted industrial policy. To truly change the game in favor of stable economic growth and shared prosperity, however, more is needed. The United States needs a comprehensive industrial strategy focused on domestic manufacturing that puts it at the forefront of the global economic transition and leaves its workers better off. And it needs to do this while demanding the creation of high-quality jobs across construction, production, operation, and maintenance. This would make climate action more politically enduring, given the benefit that such policy would have on both the climate and the livelihoods of those workers engaged in producing the goods needed to facilitate the economy's transition.

## THE REEMERGENCE OF A U.S. INDUSTRIAL POLICY

Before digging into how and why there must be job-quality connections to green industrial policy, let us take a moment to explore the United States' recent foray into industrial policy. First, it's important to differentiate industrial policy and industrial decarbonization. The former applies to a method of policy making that utilizes the public sector to directly support specific industries, whether through trade, investment, or research and development. The latter is an effort, either through public policy or private investment, to eradicate emissions from heavy industry. Work at the federal level has happened on both industrial policy and industrial decarbonization, but the two have yet to truly overlap and create synergy. This gap can and should be bridged.

### Federal Actions Stimulated a Manufacturing Renaissance in the United States

The industrial policy undertaken during the Biden administration focused on semiconductors, electric vehicles, hydrogen, and industries associated with clean energy production. The CHIPS Act is the clearest example, with dedicated grants going to companies to build semiconductor manufacturing facilities in the U.S. alongside public investment in R&D and the creation of tax subsidies to support continued production of domestically made semiconductors (CHIPS and Science Act 2022).

This resulted in the United States reemerging in the global supply chain for semiconductors and, most importantly, now having the capacity to at least partially supply itself to avoid another supply chain disaster such as was seen during COVID-19 (WhiteHouse.gov 2021). From 2020 through the end of 2023, the United States had 80 new semiconductor manufacturing projects announced. This has led to a tripling of domestic semiconductor production and a massive increase of domestic capital expenditures, with the United States securing more than 28% of global capital expenditures, an estimated $646 billion (Varadarajan et al. 2024).

All of this investment went to an industry that had gone from 100% market share in the 1970s to a low point of 8% recently and was projected to decline further if not for the passage of the CHIPS Act (Elkus 2024). Now, the United States is set to grow its share of global capacity to 14% by 2032 and will see the biggest increase in semiconductor capacity of any region in the world, with a 203% increase (Varadarajan et al. 2024). The policies and investment of the CHIPS Act were successful in directly reviving a domestic industry critical to the future of the global economy—and the domestic steel industry is in need of a similar strategy.

## Initial Investments Had Begun Incentivizing Heavy Industry to Decarbonize

The Infrastructure Investment and Jobs Act (IIJA) and IRA sparked a boom in manufacturing and clean energy and include several policies to decarbonize heavy industry. The U.S. Department of Energy's Office of Clean Energy Demonstrations awarded $1.5 billion for six iron and steel decarbonization projects in March 2024, funded by the two acts. The IRA provided grants to manufacturing facilities to decarbonize, tax incentives to support the production of key goods across the clean-tech supply chain, and tax credits to support the consumption of those products.

Alongside this, the IIJA's $1 trillion investment in U.S. infrastructure included a significant expansion of Buy America, which directed agencies to prioritize domestically produced materials for federally funded projects (Shmavonian 2023). These policies quickly started showing a major impact. Manufacturing construction spending hit its highest point in U.S. history in 2024, hovering near $240 billion, when it was just $80 billion three years ago (Richter 2024). New facilities building EV batteries and solar cells, processing critical minerals, and electrolyzing green hydrogen were taking shape across the country (Bermel et al. 2024). This had begun to remake our economy, establishing the U.S. as a leader in clean energy global supply chains and rebuilding a sputtering manufacturing sector. It created a foundation to ensure lasting U.S. industrial competitiveness for decades.

The Biden–Harris administration also invested significantly in public procurement of building materials that have low embodied carbon emissions and are produced in the United States, and the administration launched the Federal Buy Clean Initiative in 2021 (U.S. Office of the Federal Chief Sustainability Officer, no date). The IRA then provided $4.5 billion to the General Services Administration (GSA), the Environmental Protection Agency, and the Department of Transportation to identify and procure clean construction materials for federally funded construction projects. Combined, these efforts led to the federal government's first "Buy Clean" standard for low-carbon iron and steel. It was issued by the GSA, which manages federal building and construction projects (General Services Administration 2023). Several states have also implemented Buy Clean programs, and in 2023, the Biden–Harris administration launched the Federal–State Buy Clean Partnership in which 13 states committed to prioritize efforts that support the procurement of low-carbon construction materials (U.S. Office of the Federal Chief Sustainability Officer, no date).

A sad postscript to these successes is that the Trump administration and Congress went to work immediately eviscerating most of the core policies associated with the Biden clean industrial policy. As a result, much of the growth and expansion of manufacturing has—at best—slowed and possibly been reversed.

## WHAT ABOUT JOB QUALITY?
### Job Quality Actions in the Biden Administration
Thanks to Manchin, the industrial policies included in the IRA (as with the IIJA and CHIPS Act) did not explicitly include incentives or mandates to ensure that new manufacturing jobs created through the administration's industry policy actions would be high quality. The Biden administration did, however, attempt to use its authority to fill this gap by encouraging job quality standards, workforce training, and contractual agreements in many of its industrial policy investments. The Department of Energy established guidance that nearly all grants and loans from the agency that came from IIJA and IRA requires applicants to submit a community benefit plan (CBP) (U.S. Department of Energy, no date). The CBP is built from a tool long used by the labor movement and community action organizations, referred to as a community benefits agreement. As described by PolicyLink, "the [community benefits agreement] is a legally enforceable contract between community based organizations—often including local labor unions—and the developer of a project." (PolicyLink) These agreements create a productive foundation for ensuring a project truly provides benefits to the people who work in the facility and live nearby.

Public support is given for a project in exchange for commitments that can and often include direct investments in the community and current and future workforce. Notably, those commitments can incorporate agreements that address the quality of the jobs created by the project. On the construction side, this often includes a project labor agreement. For production, operation, and maintenance, this can mean a labor peace agreement—effectively a legally binding agreement that the employer will be neutral during union organizing and the union will refrain from economically disadvantageous actions like work stoppages–or other explicit commitments toward signing a collective bargaining agreement.

The Department of Energy's CBP guidance was the first of its kind at the national level, and it includes recommendations in its community and labor engagement principle to work with unions explicitly—mentioning project labor, labor peace, and collective bargaining agreements by name. The guidance did not mandate explicit cooperation with unions, so not every announcement and investment involved a good story of a unionized workforce.

Importantly, the attention the Biden administration placed on the creation of quality jobs was demonstrated well by the breadth of agencies involved in the effort. The CHIPS Act investments, for example, led to a remarkable collaboration between the Departments of Commerce and Labor to produce "Good Jobs Principles"—eight principles of job quality to guide the two departments, along with state and local governments and the private sector who plan to collaborate or request funding. These principles laid the pathway for spending the tens of billions from the CHIPS Act and resulted in positive, pro-union success stories. Combined with the efforts by the Department of Energy, a hopeful painting of job quality in manufacturing began to become clearer. This is summed up well by Aurelia Glass and Karla Walter from the Center for American Progress:

> These job quality measures represent the first governmentwide effort since the 1960s to ensure that federally supported manufacturing jobs adhere to higher standards. Agencies have begun to successfully extract commitments from manufacturers receiving funding. While the federal government should go much further to set high and consistent manufacturing job quality standards, several projects selected to receive federal grants have made firm commitments to creating good jobs. For example, Talon Metals, which previously established a workforce development partnership in Minnesota with the United Steelworkers (USW), received $114 million in funding for a minerals processing plant in North Dakota. Century Aluminum expects to employ 1,000 union workers represented by the USW for a green aluminum smelter selected as part of the DOE's Industrial Demonstrations Program. As part of the same program, the DOE also selected a union project for expansion in Middletown, Ohio, where Cleveland-Cliffs will receive up to $500 million to replace an existing blast furnace, creating 2,500 jobs represented by the International Association of Machinists and Aerospace Workers. Semiconductor manufacturer Micron received $6.1 billion to build plants in New York and Idaho after agreeing to meet with the Communications Workers of America (CWA), which represents some semiconductor manufacturing workers, to discuss a labor peace agreement. However, the government should continue to encourage recipients to achieve final agreements with worker representatives. (Glass and Walter 2024)

## WHAT NEEDS TO CHANGE?

If we are to make green industrial policy work for working people, then structural change must come in three ways. First, working people need to use the tools currently at their disposal to exercise their power and incorporate a vision for better climate jobs into their labor actions. Along with this, elected officials must use their soft power to support working people and push corporations to better share their success with their workers. Second, federal policy must strengthen and expand the tools available to working people to rebalance the power structure between workers and employers. Third, industrial policy needs to continue apace but with clear and specific labor standards attached to it.

## Leveraging Union Organizing Power

Working people should use their leverage to demand better working conditions and a path forward for their jobs in an economy that addresses the climate crisis—especially at globally competitive manufacturing facilities. The United Auto Workers (UAW) gave us an example of this leadership when negotiating with the Big Three domestic automakers in 2023. Alongside core demands about wage justice, retirement security, and worker safety, they successfully demanded that Ford, General Motors, and Stellantis committed to direct investments in EVs and their associated batteries in the United States (United Auto Workers, no date). They also negotiated neutrality agreements with the companies on organizing efforts at their joint ventures that currently manufacture batteries for EVs. As a result, UAW has successfully organized workers at GM's Ultium facility (United Auto Workers 2024) in Ohio and moved toward an election at Ford's BlueOval plant in Kentucky (Boudette 2025). Having the foresight to see the changes in their industry, UAW used its leverage and power to secure a foothold for members now and in the future.

Organizing new companies will also be necessary, and we've seen some action and victories surface over the past few years. One of the more potent victories was the United Steelworkers (USW) organizing effort at Blue Bird Corp, an electric bus manufacturer in Georgia (United Steelworkers 2023). Pushing for better wages and working conditions, the workers at the facility—over 1,500 of them—recognized the impact of federal funding and attention and used not only their leverage but also the leverage of allies in the federal government. President Biden, his administration officials, Senators Warnock and Ossoff (Georgia), and several elected officials weighed in to support the organizing drive. This critical utilization of soft power—built on top of the mandate from the multibillion-dollar grant from the Environmental Protection Agency (2024) that funds cannot be used to fight worker organizing efforts—added leverage to the workers and significantly tamped down anti-union actions by the companies. If elected officials truly want to represent working people, there are few better ways to use their power and platform. It is stated well by Karla Walter and Sachin Shiva:

> By using the bully pulpit—a public official's ability to gain attention and sway key actors through public speech and private convenings because of their prominent position—officials are increasing workers' confidence that their demands will be

respected. At the same time, they are leveraging the bully pulpit to remind corporations that the government will hold lawbreakers accountable. Yet public officials and policymakers at all levels can do more to advocate forcefully for workers throughout the entire span of a unionization and contract bargaining process, particularly when corporations are undermining worker efforts or receiving financial incentives from the public. (Walter and Shiva 2024)

## Mandate Better Green Industrial Policy

The United States must continue its green industrial policy to tackle the climate crisis equitably. Structural changes are necessary for the retention and creation of quality jobs in manufacturing. But that's not quite enough. First, there needs to be a national commitment to a comprehensive strategy that includes a robust and consistent research and development apparatus coordinated across the federal government and the private sector, along with targeted economic planning; expansive public investment in support of both the production and demand of goods; standards to mandate high-quality job creation; and a muscular trade enforcement regime.

Second, green industrial policy often rightfully incorporates significant direct public investment. That investment should be tied to standards that ensure that the jobs created in the production of goods are high-quality union jobs. The IRA succeeded in doing this for construction jobs by tying many of the incentives included in the bill to key labor standards such as prevailing wages and registered apprenticeships. For manufacturing jobs, these standards often do not apply, though research has shown that a manufacturing prevailing wage could be applied in sectors key to the budding clean economy (Glass, Madland, and Walter 2022).

In addition to prevailing wages, the most potent labor standard for manufacturing workers are the aforementioned labor peace agreements. These provisions can and should be established as a condition for federal funding. This would not only ensure project continuity and protect public investments against labor actions but also lead to better labor practices and respect for working peoples' rights.

Third, working people have too often borne the brunt of economic and technological change through loss of jobs and livelihoods. Examples can be seen in the coal and tobacco industries, as well as the manufacturing industries mentioned above. Regardless of whether the transition in industry is merited (such as with coal and tobacco), the people and communities who rely on those jobs should not be left with a more difficult economic situation. A structure of fairness for workers and communities must be built into a successful green industrial policy. Doing so would require direct investment and guidance from those workers and communities. César F. Rosado Marzán offers the idea for "Fair Transition Funds":

> These funds, like Taft–Hartley funds, would be co-managed by labor and management representatives. Green, profitable firms benefiting from industrial policy—e.g., the Teslas, Rivians, and the

Lion Electrics of the US—should play a principal role in funding them. The funds could institute supplementary unemployment benefits, retraining, and job matching programs— active labor market plans—to help move workers from non-green to green jobs, and thus better guarantee them as being "winners" of the green transition. US workers would be more apt to perceive union membership as important if unions can provide transition benefits to their members. Fair Transition Funds would thus provide a visible and influential institutional role for unions to lead in the green transition. (Rosado Marzán 2024)

## Structurally Change Federal Labor Laws

The system of labor laws is often referred to as broken because the scales are tilted so heavily toward employers. As labor activist Brett Banditelli put it, "Labor law isn't broken. It was designed this way" (Banditelli 2025). The law allows companies to drag out union election processes and make their workers sit in anti-union meetings. Much more egregiously, a worker who was wrongfully laid off can sometimes wait months or years for reinstatement and restitution, and states can institute supposed right-to-work laws, allowed by federal law. The deck is stacked purposefully. Federal labor law doesn't just need to be fixed. It needs to be overhauled. Passing the PRO Act would go a long way to rebalancing the scale, by speeding up elections, banning captive-audience meetings, barring right-to-work laws, and providing immediate relief for workers wrongfully discharged, among many other provisions (McNicholas, Poydock, and Rhinehart 2021). Additionally, the federal government should do more to allow states to be the laboratory of democracies—and recognizing the difference of industries from state to state—by passing the Public Service Freedom to Negotiate Act, which would allow states to write and administer their own labor laws.

Going further, the federal government should be putting working people in more positions of power across the economy, starting in key industries associated with green industrial policy. It can do that by establishing tripartite wage boards— collaborative bodies that include representatives from labor, management, and the government that set industry-specific wage standards, ensuring fair compensation and improving working conditions. This type of action helped lift up unions and gave workers a voice in Puerto Rico in the mid-20th century as manufacturing grew rapidly on the island. Notably, "centralized bargaining came through wage boards and institutional roles for unions to bargain for employee benefits at the plant level (Rosado Marzán 2024).

On top of this, Senator Elizabeth Warren's Accountable Capitalism Act (2020) includes a provision that would require major corporations to allow their workers to elect 40% of the membership of the board. She calls it "a crucial tool for giving millions of workers more control over corporate decisions on everything from wages and benefits to outsourcing and long-term investments" (Warren 2019).

## CONCLUSION

While recent federal initiatives such as the IRA, the CHIPS Act, and the IIJA showed success by spurring significant investments in clean technology and industrial revitalization, gaps remain—particularly in ensuring the creation of high-quality manufacturing jobs with fair wages and union representation. Green industrial policy, when implemented without sufficient labor standards, risks replicating the same inequities that have plagued U.S. workers for decades.

To address these challenges, a more comprehensive approach is required—one that explicitly ties public investment to labor standards, ensures a fair transition for displaced workers, and strengthens the tools available for union organizing. Historical evidence and recent case studies highlight the critical role of unions in improving job quality, and robust federal support can help restore the manufacturing sector as a cornerstone of middle-class prosperity. Moreover, leveraging existing policy instruments such as community benefit agreements, passing new laws such as the PRO Act, and establishing tripartite wage boards can help rebalance the power dynamics between employers and workers.

Ultimately, a successful green industrial policy must go beyond reducing carbon emissions. It should aim to build a resilient economy where both environmental sustainability and social equity are prioritized. Only by embedding strong labor standards into the core of industrial policy can the United States create a sustainable and inclusive green economy that works for all.

Successful policies should persist regardless of the actions of the federal government. However, the impetus for this cresting change—an administration and Congress dedicated to high-quality union jobs across all sectors and industries—is now gone thanks to the election of Donald Trump and an anti-union and anti-climate-action Congress. In the short term, hope for future changes now rests in state and local governments and in the hands of workers themselves. In the medium and long-term, collective work can leverage significant political and policy changes at the national level.

## ACKNOWLEDGMENTS

The author would like to thank Jamie Friedman and Trevor Sutton for previous collaborations that greatly supported ideas included in this essay, as well as Ryan Mulholland and Tara Williams for exceptional review.

## ENDNOTES

1. The following two subsections are part of a recent paper published by the author and his co-author Jamie Friedman, "The Next Frontier in American Industrial Policy: Saving the Steel Industry by Decarbonizing It" (https://tinyurl.com/kvdchrzr).

# REFERENCES

Accountable Capitalism Act. 2020. S.3215. https://tinyurl.com/36seffrn

Baker, Dean. 2024. "Manufacturing Jobs: Unions Made Them Good, Not the Factories." Center for Economic and Policy Research. August 4. https://tinyurl.com/3aemjra4

Banditelli, Brett. 2025. BlueSky thread. https://tinyurl.com/37t7hwhw

Barbaro, Michael, and Dan Kaufman. 2024. "How NAFTA Broke American Politics." *New York Times* "The Daily" podcast. October 8. https://tinyurl.com/yck7ne9r

Bermel, Lily, Ryan Cummings, Brian Deese, Michael Delgado, Leandra English, Yeric Garcia, Hannah Hess, Trevor Houser, Anna Pasnau, and Harold Tavarez. 2024. "Clean Investment Monitor: Tallying the Two-Year Impact of the Inflation Reduction Act." Rhodium Group and MIT CEEPR. August 7. https://tinyurl.com/yvnbjxuu

Boudette, Neil. 2025. "U.A.W. Seeks Union Election at Ford Battery Joint Venture in Kentucky." *New York Times*. January 8. https://tinyurl.com/33pfwyzj

Brazil Presidency. 2024. "Brazil Launches New Industrial Policy with Development Goals and Measures Up to 2033." January 26. https://tinyurl.com/3nu3tnuz

Build Back Better Act. 2021. Congressional Record—House. November 18. https://tinyurl.com/5es3cxed

Cabinet Secretariat of Japan. 2023. "The Basic Policy for the Realization of GX—A Roadmap for the Next 10 Years." February. https://tinyurl.com/32a3syav

CHIPS and Science Act. 2022. Public Law 117-167. https://tinyurl.com/3szz43en

Cornell School of Industrial and Labor Relations. No date. "What Is a Labor Peace Agreement Under MRTA?" https://tinyurl.com/bd6xrwxn

Davenport, Coral, Lisa Friedman, and Hiroko Tabuchi. 2022. "Manchin, Playing to the Home Crowd, Is Fighting Electric Cars to the End." *New York Times*. July 12. https://tinyurl.com/3e8cazj7

Elkus Jr., Richard. 2024. "Strategy for the United States to Regain Its Position in Semiconductor Manufacturing." Center for Strategic and International Studies. February 13. https://tinyurl.com/23y83cp7

Energy Innovation and Silvio Marcacci. 2022. "The Inflation Reduction Act Is the Most Important Climate Action in U.S. History." August 2. https://tinyurl.com/2vx95h5f

European Commission. No date. "The Green Deal Industrial Plan: Putting Europe's Net-Zero Industry in the Lead." https://tinyurl.com/yzbskter

Hill, Andrew T. 2015. "First Bank of the United States: 1791–1811." Federal Reserve History. December 4. https://tinyurl.com/t9n8drue

Glass, Aurelia, David Madland, and Karla Walter. 2022. "Prevailing Wages Can Build Good Jobs into America's Electric Vehicle Industry." Center for American Progress. July 6. https://tinyurl.com/mrxe55tc

Glass, Aurelia, and Karla Walter. 2024. "Four Lessons on Creating Good Manufacturing Jobs Through the Biden–Harris Administration's Industrial Investments." Center for American Progress. October 7. https://tinyurl.com/mm9ufk8x

General Services Administration. 2023. "GSA Pilots Buy Clean Inflation Reduction Act Requirements for Low-Embodied Carbon Construction Materials." May 16. https://tinyurl.com/nbertw4a

Gold, Stephen. 2016. "Manufacturing's Economic Impact: So Much Bigger Than We Think." *IndustryWeek*. February 16. https://tinyurl.com/yers9utr

Jigang, Wei. 2020. "China's Industrial Policy: Evolution and Experience." United Nations Conference on Trade and Development. July. https://tinyurl.com/mtrrzr5u

MAPI Foundation. 2016. "Infographic: The New Model of Manufacturing's Multiplier Effect."

McNicholas, Celine, Margaret Poydock, and Lynn Rhinehart. 2021. "How the PRO Act Restores Workers' Right to Unionize." Economic Policy Institute. February 4. https://tinyurl.com/2jjp977k

Milman, Oliver. 2021. "Elon Musk Slams Biden's Build Back Better Bill and its Electric Car Incentives. *The Guardian.* December 8. https://tinyurl.com/h4juuwkn

Mishel, Lawrence. 2018. "Manufacturing Still Provides a Pay Advantage, but Outsourcing Is Eroding It." Economic Policy Institute. March 12. https://tinyurl.com/4uk45xva

Mishel, Lawrence. 2021. "The Enormous Impact of Eroded Collective Bargaining on Wages." Economic Policy Institute. April 8. https://tinyurl.com/4r73yu2n

National Association of Manufacturers. 2019. "State Manufacturing Data."

O'Donovan, Caroline. 2021. "Musk Slams Union Drive in Email to Employees." *BuzzFeed News.* February 24. https://tinyurl.com/56dch2t4

Owen, Geoffrey. 2012. "Industrial Policy in Europe Since the Second World War: What Has Been Learnt?" ECIPE Occasional Paper No. 1/2012. European Centre for International and Political Economy. https://tinyurl.com/ha724y3r

Policylink.org. No date. "Community Benefits Agreements." https://tinyurl.com/bdtz373s

Richter, Wolf. 2024. "Construction Spending Squeaks to a Record Amid a Boom in Factory Spending." *Wolf Street.* September 3. https://tinyurl.com/2hxxf5tz

Romm, Tony, Jeff Stein, Rachel Roubein, and Maxine Joselow. 2022. "Manchin Says He Has Reached a Deal with Democrats on Economy and Climate Bill." *Washington Post.* July 22. https://tinyurl.com/y9s6sap2

Rosado Marzán, César F. 2024. "Fair Transition Funds, Employer Neutrality, and Card Checks: How Industrial Policy Could Relaunch Labor Unions in the United States." In *Industrial Policy 2025: Bringing the State Back In (Again).* Roosevelt Institute. https://tinyurl.com/yc8nd2yy

Salazar X., José Manuel, Irmgard Nübler, Richard Kozul-Wright, and International Labour Office. 2014. "Industrial Policy, Productive Transformation and Jobs: Theory, History, and Practice." International Labour Organization. https://tinyurl.com/yhm9p98r

Sarel, Michael. 1996. "Growth in East Asia: What We Can and What We Cannot Infer." International Monetary Fund. September. https://tinyurl.com/pc2f5awy

Shmavonian, Livia. 2023. "Biden–Harris Administration Releases Final Guidance to Bolster American-Made Goods in Federal Infrastructure Projects." Forth. August 14. https://tinyurl.com/2yn8kdtd

Toyota Newsroom. No date. "Toyota Motor Manufacturing, West Virginia" https://tinyurl.com/573rsn9c

Tucker, Todd. 2024. *Industrial Policy 2025: Bringing the State Back In (Again).* Roosevelt Institute. https://tinyurl.com/yc8nd2yy

United Auto Workers. No date. "Backgrounder on Big Three Bargaining." https://tinyurl.com/f28a5fjr

United Auto Workers. 2024. "UAW Members at Ultium Cells Reach Industry-Defining Tentative Agreement." June 10. https://tinyurl.com/5a7k54mc

United Steelworkers. 2023. "Union Yes at Blue Bird: Workers Achieve Major Organizing Victory." September 11. https://tinyurl.com/535spewz

U.S. Department of Energy. No date. "Community Benefits Plans Overview." https://tinyurl.com/5eajdz7v

U.S. Environmental Protection Agency. 2024. "Biden–Harris Administration Announces Nearly $160 Million in Grants to Support Clean U.S. Infrastructure." July 16. https://tinyurl.com/mr2khct4

U.S. Office of the Federal Chief Sustainability Office, Council on Environmental Quality. No date. "Federal Buy Clean Initiative." https://tinyurl.com/26u32r38

Varadarajan, Ras, Jacob Koch-Weser, Chris Richard, Joseph Fitzgerald, Jaskaran Singh, Mary Thornton, Robert Casanova, and David Isaacs. 2024. "Emerging Resilience in the Semiconductor Supply Chain." Semiconductor Industry Association. May. https://tinyurl.com/4kp3nsdh

Walter, Karla, and Sachin Shiva. 2024. "Public Officials Should Use Their Bully Pulpit to Support Worker Organizing and Bargaining." Center for American Progress. July 22. https://tinyurl.com/jp2p84cd

Warren, Elizabeth. 2019. "Empowering American Workers and Raising Wages." Warren for Senate website. https://tinyurl.com/bdfrpaa2

WhiteHouse.gov. 2021. "Why the Pandemic Has Disrupted Supply Chains." https://www.whitehouse.gov/cea/written-materials/2021/06/17/why-the-pandemic-has-disrupted-supply-chains/

WhiteHouse.gov. 2024. "Fact Sheet: Two Years After the CHIPS and Science Act, Biden—Harris Administration Celebrates Historic Achievements in Bringing Semiconductor Supply Chains Home, Creating Jobs, Supporting Innovation, and Protecting National Security." August 9. https://tinyurl.com/34b8h28r

Williams, Mike, and Trevor Sutton. 2021. "Creating a Domestic U.S. supply Chain for Clean Energy Technology." Center for American Progress. October 4. https://tinyurl.com/j3w8wsfn

Wrona, Joe. 2021. "Testimony for the Senate Finance Committee Forced Labor Hearing." March 18. https://tinyurl.com/yrty9nxv

# Stronger Together: The Role of Sectoral Bargaining in Advancing a Just Transition for Autoworkers

HUNTER MOSKOWITZ
*Northeastern University*

J. MIJIN CHA
*University of California, Santa Cruz*

## Abstract

In the fall of 2023, the United Auto Workers Union (UAW) embarked on a series of strikes at the Big 3 automakers to protest wage cuts and the threat of new electric vehicle (EV) joint ventures that would not be covered under existing contracts. The strikes were ultimately successful, capping off what some labor observers called the "Hot Labor Summer." Whether the increased labor activity will successfully unionize the EV future remains to be seen. A significant challenge to UAW organizing is the outsized presence of Elon Musk and Tesla, both of whom are virulently anti-union. Tesla's EV market share set a standard for nonunion vehicle manufacturing. Though UAW leader Shawn Fein pledged to organize the entire EV industry, the task is daunting and the chances for success slim. This chapter explores the challenge of EV organizing and argues that the patchwork of nonunion and union plants makes sectorwide organizing difficult. Rather, a sectoral approach to bargaining and organizing could provide an easier route to ensuring that EV jobs are high-road jobs. The chapter discusses the possibility of sectoral organizing and offers ways to advance the idea in the United States.

## INTRODUCTION

The auto industry shift to electric vehicles (EV) is clear. EV car sales increased by 35% in 2023, and one in five cars globally sold were electric vehicles in that year (International Energy Agency 2024). This shift is necessary to reduce emissions in the transportation sector, but it also indicates an uncertain future for autoworkers. New firms entering the EV space and nonunion shops present challenges for organizing as firms locate facilities in right-to-work states, which correlate with lower wages. This chapter also shows enormous challenges to organizing and working conditions within the EV battery industry. Workers in EV manufacturing facilities face dangerous exposure to chemicals and fires, and Occupational Safety and Health Administration (OSHA) violations have proliferated across these plants.

There is also uncertainty around future growth potential for the sector, especially given the new administration's elimination of the EV tax credit in President Trump's "Big Beautiful Bill" and the impact of tariffs on suppliers for the industry (Elliot 2025; Iacurci 2025). Moreover, the inability to secure a majority on the National Labor Relations Board (NLRB) until 2026 before a change of administration has left workers' rights as vulnerable as EV support (Hubbard 2024). While the NLRB has made strides to make union organizing easier through the *Cemex* decision, which aims to limit employer anti-union activities, a Trump-controlled board could roll back protections and leave workers uniquely vulnerable to anti-union tactics (Mallon 2024).

Given the political and policy uncertainty, we argue that a shop-by-shop organizing model—the current dominant model of organizing in the United States—will prove too resource-intensive and time-consuming to ensure union saturation in the EV sector. Building on work by Dupuis et al. (2024), we argue that more structural reforms are required to protect workers in the EV transition. Specifically, we argue that sectoral bargaining, where working conditions are negotiated for an entire sector, rather than shop by shop, would more directly advance a just transition for U.S. autoworkers.

We begin by reviewing relevant literature on the EV transition and why a "just transition" is desirable. We then discuss the shop-by-shop approach, and the UAW's recent efforts and challenges before moving to a discussion of various pathways to sectoral bargaining. While these types of reforms have had limited success previously, we argue that the recent election calls for bold proposals that deliver real benefits to workers and the working class. Rather than advancing workers' rights through incentivizing private capital, we argue that directly providing workers' rights is the type of proposal needed to meet this challenging political moment.

## JUST TRANSITION: HOW WORKERS CAN BE PROTECTED DURING SECTORAL TRANSITIONS

The concept of a just transition provides a useful framework for analyzing the emergence of the EV industry and how the needs of workers can be prioritized within economic transformations. The idea of a just transition arises from environmental justice communities and Tony Mazzochi, a labor leader in the Oil, Chemical and Atomic Workers International Union, who argued for a "superfund" for workers in dangerous and environmentally harmful industries that could pay for a social safety net as these industries declined (Leopold 2007). The threat of the intensifying climate crisis has led to just transition becoming an important framework for unions, activists, and policy makers who argue that climate programs and legislations must protect workers in the fossil fuel industry and create a green energy economy with strong labor standards. The language has widespread usage, ranging from the rallying cry during the 2023 UAW strike to the call for "a fair and just transition for all communities and workers" to the introduction of the legislation Green New Deal (United Auto Workers 2023b).

The idea of a just transition works to move past the false tension of "labor versus the environment," which scholars argue has long dominated the relationship between industrial unions and environmental groups within the public eye (Morgenstern, Pizer, and Shi 2002; Räthzel and Uzzell 2013). In a "labor versus the environment" approach, the interests of unions and environmentalists are seen as fundamentally opposed in a zero-sum game. Increased economic production will harm the environment and environmentally conscious policies will cause job loss. For workers, the idea of a "job blackmail" has been particularly important, where corporations push unions to oppose any environmental regulation justified as a threat to employment (Kazis and Grossman 1982).

Just transition offers an alternative pathway as an effort to protect communities from pollution and environmental injustice and to protect workers from the loss of jobs and compensation during industrial decline. Worker and community protection goes beyond only a funding stream or a social safety net and includes organizing workers in sectors in the clean energy economy and creating a wider acceptance of the necessity for economic transition to become a solution to the climate crisis (Morena, Krause, and Stevis 2019). If just transition embraces a "four+ pillars" approach with governmental support, guaranteed funding streams, diverse coalitions, economic diversification, and policies that look to disrupt the status quo, it has the potential to provide a remaking of the economy and society that moves away from extraction (Cha 2024). Scholars have pushed the concept of just transition to become a societal transformation that includes expanding energy access, meeting global inequalities and prioritizing the Global South, and reforming society's relationship with technology and consumption (Newell and Mulvaney 2020; Satheesh 2020).

Previous research examined the approach of autoworker unions to environmental concerns and the climate crisis. Unions in Canada have historically professed the rhetoric of social unionism, seeing labor organizations as vehicles for expanding progressive politics, but this has been weakened in the auto sector because unions tend to ignore the impact of automobility and carbon emissions from manufacturing on the environment (Hrynyshyn and Ross 2011). Canadian autoworkers prioritized jobs first, not incorporating environmental concerns into bargaining, and short-term issues of job maintenance have been prioritized over the climate crisis. Similarly, studies of the German auto sector have found that the rhetoric of green jobs appeals best to workers, and while leaders and environmentalists within unions advocate for ecological modernism, most workers express a limited working-class environmentalism with employment issues as the priority (Allan and Robinson 2022).

## THE 2023 UAW STRIKE AND THE DEMAND FOR A JUST TRANSITION

The UAW strike of 2023 encapsulated many of the ideals of just transition as it emerged to address an unjust transition for workers in the move away from internal combustion engines to EVs. On September 15, 2023, the UAW began a six-week strike, and while higher wages with cost-of-living adjustments and ending tiered structures were important parts of the efforts, for the first time, the strike explicitly addressed union contract coverage of the EV industry. The UAW built upon this language of just transition declaring that "[t]he UAW supports the transition to a clean auto industry" and "battery jobs will eventually replace existing engine and transmission jobs, with plants like Belvidere Assembly already closing" (United Auto Workers 2023). Unlike unions in other industries, the UAW has not shied away from openly acknowledging job loss. While the creation of new jobs at Ultium Cells and other battery plants has allowed them to take this approach more easily, it does show a willingness by the union to acknowledge that job changes, including losses, will occur, so there is a need to proactively plan for the clean energy economy. The 2023 strike became part of this approach, using the strike itself to create better conditions for this emerging industry.

The UAW declared in press statements that "electric vehicle jobs must be good jobs" and that "the EV transition is a historic opportunity to raise autoworker standards instead of lowering them" (United Auto Workers, no date). As one UAW press release declared in October of 2023: "UAW Wins Just Transition at General Motors." In this way, the UAW embraced a transformational notion of just transition (Vachon 2023), one that sees a just transition as a means of reducing inequality and improving the economy and society, and not merely just replacing one job with another. The UAW also argued that "forcing workers to decide between good jobs and green jobs is a false choice" (United Auto Workers, no date). This rhetoric counters the "jobs versus the environment" discourse that has long dominated the framing of labor–environmental conflict, illustrating how just transition can emerge as a path out of economic and environmental conflict as a zero-sum game.

In previous actions, the UAW struck one automaker and came to an agreement that would set the pattern for the other Big 3—Ford, General Motors, and Stellantis (formerly Chrysler). In this "stand-up" strike, informed by the new reformist leadership of the union under the direction of its president, Shawn Fain, the UAW initiated last-minute walkouts at key plants across the country (United Auto Workers 2023). This approach generated media attention, preserved the strike fund, and put intense stress on parts of the manufacturing process (Lichtenstein 2024). Workers won general wage increases, forms of cost-of-living increases, and many improvements for those in lower-tiered or temporary positions. And, most notably, Ultium Cells, GM's battery manufacturer, would now be included within the GM master agreement, and the union won agreements at all the Big 3 automakers for inclusion of EV workers under their contracts. However, workers at Ultium still faced a lower wage than the typical UAW worker (Krisher and Chandler 2024; United Auto Workers 2024).

The UAW strike demonstrates the power of the language of just transition and its ability to rally workers to fight for their survival within changing economic conditions. Yet without full coverage or equal benefits and uncertainty about the future of EV investment despite past promises by automakers, the actual implementation of a just transition remains unfulfilled.

## SHOP-BY-SHOP ORGANIZING AND ENTERPRISE BARGAINING

As unions have sought to organize the EV industry in the United States, they have often taken a shop-by-shop approach to organizing, where unions and employers contest union representation at each plant individually. This approach defines the labor relations system in the United States of enterprise bargaining, a decentralized approach in which unions secure specific contracts for each plant. Harvard's "Clean Slate Report" has pointed to three major challenges of the enterprise level of bargaining model: (1) a shop-by-shop approach excludes millions and millions of workers from bargaining which, in turn, exacerbates social inequalities; (2) the approach incentivizes anti-union activity because of wage competition; and (3) it fails to meet the flexible workplace of the global economy and its decentralized structure (Block and Sachs 2020) Behrens, Colvin, Dorigatti, and Pekarek (2020) found that workplace conflict resolution in the United States is defined by "fragmentation, decentralization, privatization, and individualization," with little collective voice of workers, even in unionized disputes.

Since the 1980s, across the world, countries have shifted from national and sectoral bargaining toward plant- and enterprise-level bargaining. An increase in bargaining power of corporations, the reorganization of work arrangements, corporate decentralization, and declining worker solidarity have all been offered as explanations for these trends (Katz 1993). Australia has particularly seen a push from a more cohesive system to one focused on the workplace, where bargaining has become longer and less efficient, and while including more local union participation, also created acrimony within the system (Townsend, Wilkinson, and Burgess 2013).

In general, as Bronfenbrenner et al. (1998) have pointed out, up until the late 1990s, there was limited research on union organizing within the United States, and the work that did exist typically emphasized quantitative trends in worker preference and included few studies of successful campaign tactics.

Since that period, a flurry of "union revitalization" literature has emerged, focusing on how a resurgence of the labor movement can be built through social unionism, involving communities outside the workplace, and creating transnational connections across supply chains (Doellgast, Bidwell and Colvin 2021). An examination of the video game industry found increased union support in general when workers were asked about an industrywide approach rather than an enterprise one (Weststar and Legault 2017). In sum, much of this literature has looked for a way out of the shop-by-shop approach, seeing struggles as solely between an employer and a union confined to the workplace as a part of union decline. However, the success of the union, Workers United, in organizing over 500 Starbucks stores has pushed more focus on the shop-by-shop approach (Burt 2024; Kelly 2024). Some scholars have argued that Starbucks and the Amazon Labor Union may offer a model for future shop-by-shop organizing, particularly with the important role of worker–organizers and gaining unprecedented news coverage for a union campaign (Logan 2022, 2023). However, monumental differences exist between the service sector and manufacturing, and the UAW has a long history of manufacturing union drives that need further explanation to understand its shop-by-shop approach.

Over the past decade, the UAW has taken a shop-by-shop approach to organizing. In examining union elections in the automaking industry (Table 1), the UAW has filed for 19 elections at auto manufacturing facilities since 2014, losing 10 of these elections. While the UAW has more recently invested in organizing and won more elections recently, it has still lost significant large organizing drives such as the Mercedes plant in Tuscaloosa in May 2024, which is not included in these data because the case is still open at the NLRB. Certainly, there is some optimism that workers' gains can occur through shop-by-shop organizing at the UAW, but organizing through this approach is an uphill battle. For instance, the Volkswagen plant in Chattanooga has held three elections over the past decade alone. Particularly interesting in Table 1 are the suppliers and the broader car manufacturing supply chain. As these data show, the UAW attempted to capture some parts of the supply chain not covered by current contracts, such as glass, plastics, interior, and systems manufacturing, but success in this space has been limited. A broader approach may be needed when it comes to EVs to examine the supply chain around cell assembly and final battery facilities because plants that produce cathodes, anodes, computer chips, and separators within EV batteries will be important parts of the supply chain (Yang, Huang, and Lin 2022).

## NEW VENTURES: LACK OF UNION REPRESENTATION AND LOW-WAGE WORK

In its March 2019 white paper examining the potential and the challenges confronting workers within the EV industry, the UAW identified two possible negative impacts of EV implementation, one of which was "lower complexity, less labor, displaced workers" (UAW Research 2019). Put simply, assembling EVs is simpler than assembling cars fueled by gasoline. Powertrains in EVs require far fewer parts and less mechanical complexity, leading to fewer hours per workers required to assemble and fewer jobs overall.

The threat of the loss of work initially was highly motivating for the UAW and workers concerned with the impacts of the transition to EVs. In particular, a lack of organization within this industry could compound with a loss of jobs, leading to far less power for currently employed union workers. However, not all recent research agrees that such a substantial job loss will occur. A recent study by Cotterman, Fuchs, Whitefoot, and Combemale (2024) found that EVs could actually create more jobs in the near term. These results echo the UAW's rhetoric of a just transition during the 2023 UAW strikes that sees a future for autoworkers with the EV industry. The emergence of EV manufacturing jobs and, crucially, a supply chain related to the manufacturing of batteries explains why UAW has been willing to take a more militant approach than other unions facing industrial transition. It is a clear pipeline for future work that unions see as sites of potential organization.

Table 1. UAW Auto Manufacturing Elections 2014 Through 2024—At Least
300 Eligible Workers (National Labor Relations Board, 2021–2024)

| Plant | Date of election | Result | Vote |
|---|---|---|---|
| Volkswagen—Chattanooga, TN | 2/14/14 | Union rejected | 626 for, 712 against |
| Adac Plastics—Muskegon, MI | 12/18/14 | Union rejected | 152 for, 490 against |
| Westport Axle—Breinigsville, PA | 11/21/14 | Union rejected | 126 for, 169 against |
| Fuyao Glass America—Moraine, OH | 11/9/17 | Union rejected | 444 for, 868 against |
| Nissan—Canton, MS | 8/4/17 | Union rejected | 1,307 for, 2,244 against |
| Eberspaecher—Brighton, MI | 11/16/17 | Union approved | 223 for, 96 against |
| KAMAX—Lapeer, MI | 10/11/17 | Union rejected | 125 for, 152 against |
| Volkswagen—Chattanooga, TN | 6/14/19 | Union rejected | 776 for, 833 against |
| Flex-N-Gate, Detroit, MI | 6/5/19 | Union approved | 277 for, 67 against |
| Flex -N-Gate, Shelby, MI | 12/11/18 | Union approved | 159 for, 97 against |
| Thai Summit America—Howell, MI | 11/29/18 | Union rejected | 126 for, 416 against |
| Yangfeng—Riverside, MO | 10/23/19 | Union rejected | 174 for, 215 against |
| Constellium Automotive USA—Belleville, MI | 10/30/19 | Union approved | 206 for, 167 against |
| Flex-N-Gate, Chicago, IL | 10/27/20 | Union rejected | 52 for, 113 against |
| Dakkota Integrated Systems—Hazel Park, MI | 7/15/22 | Union approved | 172 for, 0 against |
| Ultium Cells—Warren, OH | 12/8/33 | Union approved | 710 for, 16 against |
| Yangfeng—Riverside, MO | 3/3/23 | Union approved | 310 for, 26 against |
| Antonlin Interiors—Howell, MI | 1/30/24 | Union approved | 238 for, 42 against |
| Volkswagen—Chattanooga, TN | 4/20/24 | Union approved | 2,628 for, 985 against |

These data included only closed cases. Excluded for not being auto manufacturing were CNH
Industrial America, TVS Supply Chain Solutions, Caterpillar Logistics, General Cable Industries,
and all elections at medical colleges, universities, and casinos.

The second challenge that the UAW presented in its 2019 white paper was "new
industry actors." While the loss of work remains to be seen, the entrance of new
ventures has proliferated across the EV industry, especially in the creation of electric
batteries. These ventures have occurred through the emergence of new firms, investment
by foreign corporations, and operations of current auto manufacturers reorganized
into new entities.

In terms of new entrants, Tesla emerged as the first and most important company
in the EV industry, particularly demonstrating the large U.S. and even global consumer
demand for electric cars. Tesla also set the standard for labor relations, engaging in
hostility toward any attempts at unionization and fostering poor working conditions.
The Center for Investigative Reporting documented unsafe conditions, injuries, and
inadequate safety enforcement and reporting at Tesla plants (Evans 2021). While
Tesla raised wages at its factories after the UAW strike, they still lag behind the wages

paid to Ford and GM workers (Naughton and Kay 2024). Tesla attempted to stymie efforts in a 2017 unionization attempt, both threatening workers engaging in concerted activity and attempting to stop their organizing by hiring them into different positions (Campbell 2019).

Tesla has also utilized the court systems to block efforts by the NLRB to hold the EV company accountable. In the past years, courts have allowed Tesla to implement workplace uniforms that barred union apparel and even supported Tesla CEO Elon Musk's right to tweet "Nothing stopping Tesla team at our car plant from voting union. Could do so tmrw if they wanted. But why pay union dues & give up stock options for nothing? Our safety record is 2X better than when the plant was UAW & everybody already gets healthcare"—during the 2017 unionization attempt, while the NLRB itself dismissed employees at a factory in Buffalo engaging in a unionization drive (National Labor Relations Board 2023; U.S. Court of Appeals for the Fifth Circuit 2023, 2024). Tesla has utilized the legal system to avoid accountability for the way it treats employees and for its tactics to stop labor organizing.

In addition to car manufacturing, other new entrants have emerged within the vehicle battery industry, including U.S.-based companies entering the space for the first time and foreign companies who have invested in U.S. manufacturing, such as SK Battery's plant in Commerce, Georgia, which employs almost 3,000 workers (Georgia Department of Economic Development 2024). Just as common as new entrants are joint ventures—partnerships typically between established car manufacturers and companies with experience producing batteries. Most prominently, Ultium Cells, a joint venture between GM and LG Energy Solutions, operates two battery plants within the United States in Spring Hill, Tennessee, and Lordstown, Ohio. In announcing the creation of Ultium Cells, GM stated it would allow a "flexible, modular approach to EV development" (General Motors Company 2024).

Importantly, building batteries in a joint venture also allowed a different approach to labor organization, and initial plants were not included in the UAW master agreement. UAW achieved the incorporation of these plants in its agreement only through the success of organizing campaigns at both plants and through the 2023 UAW strike. Yet the joint venture, which has stopped using the name Ultium, still pays workers less than their counterparts at other GM plants and has set a model for an EV battery industry emerging outside the current union structure.

A closer examination of the currently operating and planned EV battery facilities across the United States reveals a reliance on low-wage work and a hostility to union organizing. Twenty-five EV battery plants employing more than 1,000 workers are operating or currently planned in the United States as of October 2024. Although plans for some of these plants have been stalled or canceled in 2025—particularly due to the impact of tariffs—these data still demonstrate how auto companies have and continue to think about investment within the industry (Elliott 2025). Of these 25 plants, companies located 16 in right-to-work states, typically hostile to unions and unionizing. In addition, a majority also reside in public use microdata census areas, where the average wage is at least 10% below the state average wage, indicating that these plants, especially those who supply work to the factories, may rely on a low-wage workforce. In sum, companies locate EV plants in states often more hostile to unions and in localities with more low-wage workers. While these plants could have the potential to improve the lives of working people, they could also exacerbate inequalities instead of producing a just transition if union organizing campaigns do not occur.

## HEALTH AND SAFETY ISSUES

Health and safety issues arise in the production of EVs, and especially batteries, because workers face exposure to toxic chemicals. Factories without union contracts have less protection and support for workers, are less likely to provide proper safety and health equipment, and have fewer channels for workers to voice their complaints. During its organizing campaign at Ultium Cells in Lordstown, the UAW released a white paper in 2023 detailing the health and safety issues that workers faced within the plant, including high turnover, suspension of workers who refused to participate in unsafe work, and dangerous incidents such as workers being sprayed with toxic chemicals (United Auto Workers 2023a). The UAW argued that "ramping up EV production to reduce climate impacts must not result in spreading dangerous manufacturing practices to communities across the country" (United Auto Workers 2023a). As Table 2 (next page) shows, Ultium, SK Battery, and Tesla demonstrate the numerous hazards plaguing battery manufacturing. A number of fines in the fall of 2023 relate to an explosion at the plant where OSHA found 19 safety and health violations.

As Table 2 demonstrates, OSHA has fined Ultium, SK Battery, and Tesla for all manner of safety and health violations, from exposing workers to toxic chemicals to not having proper protective equipment to not implementing appropriate safety procedures and evacuation plans. The threat of chemical exposure especially seems to be a paramount challenge, one that the UAW argues can be solved only through unions who can enforce compliance. Yet without a widespread increase in union representation, many workers may be forced to face such a dangerous environment with little recourse on the shop floor. As one UAW worker stated in a testimonial about Ultium in Lordstown, "I think the general consensus of the employees there is they don't feel comfortable. They don't feel that it's safe or even producing something that is worth putting out" (United Auto Workers 2023a). The EV industry and unions will need to confront these challenges, particularly the dangers of manufacturing with these chemicals and the inequalities created by unequal safety standards across factories.

Indeed, while not recent, the health and safety violations at Tesla included incidents that ended with an amputation and fatality (Table 2). The unsafe and improper procedures and manufacturing, particularly in nonunion spaces, can have dangerous outcomes for workers. Given concerns for decades over OSHA's limited resources and low effectiveness in preventing injuries, a more comprehensive approach rather than focusing on individual plants may be warranted (Johnson, Levine, and Toffel 2023).

Table 2. OSHA Fines Against Ultium Cells and SK Battery America
(Occupational Safety and Health Administration, no date).

| Plant | Date fine was issued | Example text of violation |
|---|---|---|
| Ultium Cells—Lordstown | 6/27/24 | "The employer did not ensure to identify the intended use of the cleaning procedure" |
| Ultium Cells—Lordstown | 5/10/24 | "The employer had not corrected equipment deficiencies that were outside acceptable limits defined by process safety information" |
| SK Battery America | 3/15/24 | "The employer exposed employees to inhalation hazards from a toxic atmosphere" |
| SK Battery America | 1/3/24 | "The employer exposed employees to severe laceration/amputation hazards" |
| Ultium Cells—Lordstown | 12/12/23 | "The employer did not report within 24-hours a work-related incident resulting in in-patient hospitalization" |
| Ultium Cells—Lordstown | 10/2/23 | "Packaging employees tasked with loading magazine trays in cell assembly were exposed to tripping hazards from the scrap duct work" |
| Ultium Cells—Lordstown | 10/3/23 | "The employer did not develop and implement an emergency response plan to handle anticipated emergencies prior to commencement of emergency response operations" |
| Ultium Cells—Lordstown | 10/2/23 | "The employer did not select and require employees to use appropriate hand protection when employees' hands were exposed to hazards" |
| Ultium Cells—Lordstown | 1/13/23 | "Procedures were not developed, documented and utilized for the control of potentially hazardous energy" |
| SK Battery America | 12/8/23 | "The employer exposed employees to respiratory hazards associated with Nickel" |
| Ultium Cells—Lordstown | 11/7/22 | "The employer did not include the procedures for reporting a fire within their emergency action plan" |
| SK Battery America | 12/23/22 | "Employees were exposed to hazardous chemical/materials such as but not limited to acetonitrile, isopropyl alcohol, ethyl alcohol, and an adhesive solution" |
| Tesla—Sparkz | 2/21/21 | "An employee was changing a spool of wire on some machinery and was found unresponsive at their work station. The employee died of atherosclerotic and hypertensive cardiovascular disease while at work." |
| Tesla—Sparkz | 9/12/19 | N/A |
| Tesla—Sparkz | 11/14/18 | "An employee was trying to determine the compatibility of a lower guard for a pair of pneumatic wire cutters. The employee removed the top guard from a pair of Simonds wire cutters, Model PTX-938, due to it not being compatible with the newly designed lower guard. The employee operated the cutters and amputated his right hand's middle fingertip" |

# THE DIFFICULTY OF ORGANIZING BY ENTERPRISE

The case of the EV industry demonstrates the challenges of organizing across plants or enterprises rather than throughout the sector itself. Geographical inequalities between right-to-work states and other states present a problem for union campaigns, particularly in the South. UAW victories at Spring Hill and Volkswagen in Chattanooga demonstrate that campaigns can win, but scholars should not overlook recent setbacks, such as at Mercedes factories in Alabama (Krisher and Chandler 2024; United Auto Workers 2024). It should be noted that most of these companies, including Volkswagen and Mercedes, have unionized workforces in their home countries, but they fight unionization attempts in the United States. An enterprise-by-enterprise approach leaves many workers to face health and safety issues without protections, takes considerable time, and could strain the resources of the union itself as it faces expensive organizing campaigns.

Contract language in the Ultium Cells agreement also demonstrates how organizing plant by plant could lead to consequences across the industry. UAW's agreement with GM at Ultium contains "a most-favored nation" clause, which states that if UAW organizes other battery plants and signs an agreement that "would put Ultium Cells at a materially less favorable position on overall labor costs (i.e., base wages, healthcare, and retirement contributions)," the parties would meet and negotiate language in how to "rectify the imbalance" (United Auto Workers Local 1112, 2024). If no agreement is reached, GM has the right to renegotiate the agreement with the competitive labor costs in mind. This clause has the potential to put downward pressure on wages at Ultium Cells and throughout the industry because new entrants could undercut the wages negotiated in this contract and force lower wages across the industry if the newly bargained contracts have lower labor costs. This clause could further jeopardize the possibility of a just transition as wages sink to the level of the lowest common denominator. It also incentivizes workers in the UAW to invest less in other organizing campaigns because doing so could result in lower wages if the newly organized shops have weaker contracts. In this way, such contract language represents the problems with the enterprise model, which forces workers to think about their immediate needs instead of a more sustainable transition for the industry.

Similarly in December 2024, news outlets reported that the UAW had reached a neutrality agreement for union organizing at Rivian, but workers could unionize only if Rivian became profitable (Eidelson 2024). While organizing at companies such as Rivian would improve the prospects of wider EV unionization, it also demonstrates another fragmentation of the representation across the industry given the conditions imposed about profitability. Though workers at Ultium may be unionized, those at Rivian will not be until the company turns a profit—an uncertain prospect for a company losing hundreds of millions of dollars quarterly (Subramanian 2024). Not only does this tie workers' labor rights to their company's prospects—a difficult challenge for workers who have no control over company decision-making—it also sets another barrier for solidarity across the industry, as the prospects of Rivian workers now depend on the company's success and possibly the failure of its rivals. An organizing structure that pits the prospects of workers against each other either through most-favored nation clauses or through these types of organizing agreements may make it harder for workers to move together and fight for a true just transition. In this way, a broader approach that considers the economy as a whole is needed.

## MOVING BEYOND SHOP-BY-SHOP: THE POSSIBILITIES OF SECTORAL BARGAINING

In their paper, "A Just Transition for Auto Workers? Negotiating the Electric Vehicle Transition in Germany and North America," Dupuis et al. (2024) detail the German auto industry transition and apply lessons from that transition to U.S. and Canadian unions. Dupuis et al. argue that, owing to wage tiers, poor working conditions, safety issues, and lack of union power in the North American EV industry, unions in North America should follow the approach of German unions to ensure a just transition. Unions in Germany have structural power in the automobile industry, particularly through work councils and sectoral bargaining. The U.S. labor system must make it easier for unions to organize (especially in the South), encourage broader coordination of bargaining, mandate bargaining over investment decisions, involve unions more closely in policy making, and expand the social safety net to help transitioning workers.

A key recommendation is for collective bargaining at the sectoral level and, importantly, tying government incentives to either the plant agreeing to sectoral bargaining or at least maintaining neutrality—otherwise known as a labor peace agreement. The idea of sectoral bargaining, or variations of it, is not new to the United States. Liechtenstein details how during the Progressive Era, several states established boards to set wages and other labor standards (Lichtenstein 2022). While supervised by a state commissioner, employers and employees did most of the work, collected evidence, and bargained to reach a compromise on a minimum wage for a variety of industries. California's minimum wage for fast food workers and the nationwide Fight for $15 campaign are also examples of sectoral standards (California Department of Industrial Relations 2025).

Larry Cohen, past president of the Communication Workers of America, argues that political and economic conditions, specifically the strength of employer and corporate opposition, mean that more funding for organizing workers is not enough (Cohen 2018). Cohen notes that even when Democrats are in power, labor rights see little advancement. Given the likelihood of the NLRB turning hostile to workers in the short term, as it did during Trump's first presidential term, Cohen's argument is particularly relevant. Organizing workers is important, but the site-by-site model is too limiting and costly. The institutions meant to protect and advance workers' rights, such as the NLRB, are inadequate to protect workers across sectors. This inadequacy is well highlighted by the reality that Tesla continually violates workplace, corporate, and tax regulations, yet the fines levied seem to have no noticeable impact on the company because it continues operating as usual (Good Jobs First 2025), which tracks the fines that Tesla paid. Fine do not seem to deter Tesla or force the company to change its behaviors. The institutions meant to advocate and protect workers have been greatly weakened and even under friendly administrations are limited in how they can support building worker power. Relying on institutions or benevolent employers has failed to advance sectorwide gains for workers.

While the benefits of sectoral bargaining are clear, Veena Dubal (2022) raises concerns over advocating sectoral bargaining without also building worker power.

She cites the example of rideshare workers in California to caution that sectoral bargaining does not necessarily lead to worker empowerment. In California, a failed proposal for rideshare workers would have established a narrow set of sectoral standards but, in return, workers would, "give up any and all claims to the rights and protections granted to employees under U.S. law" (Dubal 2022). A similar proposal in New York also failed when it became clear that workers would have to give up basic protections and accept lower wages. In Massachusetts, a similar ballot measure passed in November 2024, which has been controversial, with some labor advocates declaring that it creates "company unions" (Goldstein 2024; Raymond 2024). These examples note that sectoral bargaining, in and of itself, is not enough to ensure strong worker and workplace protections. Rather, sectoral bargaining works in tandem with building worker power. Organizing combined with sectoral bargaining is a path for ensuring that workers across the EV industry have voice and safe working conditions.

Though difficult, sectoral bargaining could take a variety of forms. Similar to the Fight for $15, workers could demand a uniform wage across the sector. Public funding to the EV industry, including financial incentives, grants, and loans, could be conditioned on meeting sectoral standards for wages and workplace conditions. Firms do not have a fundamental right to public resources, and federal, state, and local authorities should condition the use of public resources more to advance policies and practices that are beneficial for workers and communities. Wage boards that are dormant could be revived to set standards across industries, similar to how they operated in the 1900s. Finally, companies could sign voluntary standards agreements. Industry-wide standards would level costs so that no one firm would be at a disadvantage by providing decent wages and working conditions.

## CONCLUSION

The second Trump administration may have succeeded in removing some of the financial incentives for EVs, but in all likelihood, the transition will continue. Given the near certainty of this change, the UAW's strikes showed the importance of not just transitioning but also transferring worker protections and wages to the emerging EV industry. Sectoral bargaining offers a way to more effectively, uniformly, and fairly ensure workers are not left behind in the transition. Though challenging, the political realities of the second Trump administration call for bold, ambitious visions that counter the dystopian ones the administration promises. Sectoral bargaining alongside building worker power offers a concrete vision for how the low-carbon economy can provide livable wages and safe working conditions for transitioning workers—an actual just transition.

## REFERENCES

Allan, Kori, and Joanna Robinson. 2022. "Working Towards a Green Job?: Autoworkers, Climate Change and the Role of Collective Identity in Union Renewal." *Journal of Industrial Relations* 64, no. 4: 585–607. http://doi.org/10.1177/00221856221088153

Behrens, Martin, Alexander J.S. Colvin, Lisa Dorigatti, and Andreas H. Pekarek. 2020. "Systems for Conflict Resolution in Comparative Perspective." *ILR Review* 73, no. 2: 312–344. https://doi.org/10.1177/0019793919870800

Block, Sharon, and Benjamin Sachs. 2020. "Clean Slate for Worker Power: Building a Just Economy and Democracy." Labor and Worklife Program, Harvard Law School. January 1.

Bronfenbrenner, Kate, Sheldon Friedman, Richard W. Hurd, Rudolph A. Oswald, and Ronald L. Seeber, eds. 1998. *Organizing to Win: New Research on Union Strategies.* ILR Press.

Burt, Kristian. 2024. "500 Starbucks Locations Have Voted to Unionize as Labor Talks Continue." CNBC. October 1. https://tinyurl.com/4ushrns9

California Department of Industrial Relations. No date. "Industrial Welfare Commission Wage Orders." https://tinyurl.com/yz8k7mt9

Campbell, Alexia Fernández. 2019. "Elon Musk broke US labor laws on Twitter." *Vox.* September 30. https://tinyurl.com/y7285dmd

Cha, J. Mijin. 2024. *A Just Transition for All : Workers and Communities for a Carbon-Free Future.* MIT Press.

Cohen, Larry. 2018. "The Time Has Come for Sectoral Bargaining." *New Labor Forum* 27, no. 3: 10–13. https://doi.org/10.1177/1095796018789908

Cotterman, Turner, Erica R. H. Fuchs, Kate S. Whitefoot, and Christophe Combemale. 2024. "The Transition to Electrified Vehicles: Evaluating the Labor Demand of Manufacturing Conventional Versus Battery Electric Vehicle Powertrains." *Energy Policy*, 188, 114064. https://doi.org/10.1016/j.enpol.2024.114064

Doellgast, Virginia, Matthew Bidwell, and Alexander J.S. Colvin. 2021. "New Directions in Employment Relations Theory: Understanding Fragmentation, Identity, and Legitimacy." *ILR Review* 74, no. 2: 555–579. https://doi.org/10.1177/0019793921993445

Dubal, Veena. "Sectoral Bargaining Reforms: Proceed With Caution." *New Labor Forum* 31, no. 1: 11–14. https://doi.org/10.1177/10957960211061723

Dupuis, Mathieu, Ian Greer, Anja Kirsch, Grzegorz Lechowski, Dongwoo Park, and Tobias Zimmermann. 2024. "A Just Transition for Auto Workers? Negotiating the Electric Vehicle Transition in Germany and North America." *ILR Review* 77, no. 5: 770–798. https://doi.org/10.1177/00197939241250001

Eidelson, Josh. 2024. "UAW Reaches Secret Deal with Rivian to Make Unionizing Easier." *BNN Bloomberg.* December 19. https://tinyurl.com/5n7tcwnb

Elliot, Rebecca F. 2025. "Trump's Trade and Tax Policies Start to Stall U.S. Battery Boom." *New York Times*. June 16. https://tinyurl.com/2js4dfkd

Evans, Will, and Alyessa Jeong Perry. 2021. "Tesla Says Its Factory Is Safer. But It Left Injuries Off the Books." *Reveal.* April 16. https://tinyurl.com/3x54fz8h

General Motors Company. 2024. "GM Reveals New Ultium Batteries and a Flexible Global Platform to Rapidly Grow Its EV Portfolio." March 4. https://tinyurl.com/3kdxpu29

Georgia Department of Economic Development. 2024. "SK Battery America Exceeds Hiring Goal, On Track to Reach 3,000 Workers." November 4. https://tinyurl.com/5n6cykrz

Goldstein, Luke. 2024. "Massachusetts Ballot Measure Criticized for Creating Gig Worker 'Company Unions'." *American Prospect.* September 23. https://tinyurl.com/5ftpynv2

Good Johs First. 2025. "Violation Tracker, Tesla Inc." https://tinyurl.com/yux4nwpx

Hrynyshyn. Derek, and Stephanie Ross. 2011. "Canadian Autoworkers and the Climate Crisis, and the Contradictions of Social Unionism." *Labor Studies Journal* 36, no. 1: 5–36. https://doi.org/10.1177/0160449X10389747

Hubbard, Kaia. 2024. "Senate Democrats Fail to Secure NLRB Majority Under Trump in Razor-Thin Vote." CBS News. December 11. https://tinyurl.com/mskrjzf7

Iacurci, Greg. 2025. "Trump Megabill Axes $7,500 EV Tax Credit After September." CNBC. July 1. https://tinyurl.com/muvsubxn

International Energy Agency. 2024. "Global EV Outlook 2024: Moving Towards Increased Affordability." April. https://tinyurl.com/ywtyp9a6

Johnson, Matthew S., David I. Levine, and Michael W. Toffel. 2023. "Improving Regulatory Effectiveness Through Better Targeting: Evidence from OSHA." *American Economic Journal: Applied Economics* 15, no. 4: 30–67. https://doi.org/10.1257/app.20200659

Katz, Harry C. "The Decentralization of Collective Bargaining: A Literature Review and Comparative Analysis." 1993. *ILR Review* 47, no. 1: 3–22. https://doi.org/10.2307/2524228

Kazis, Richard, and Richard L. Grossman. 1982. *Fear at Work: Job Blackmail, Labor and the Environment.* Pilgrim Press.

Kelly, Sara. 2024. "Message from Sara: Starbucks and Workers United Agree on Path Forward." February 27. Starbucks. https://tinyurl.com/8zuy2uvh

Krisher, Tom, and Kim Chandler. 2024. "After Decisive Loss at Alabama Mercedes Plants, Powerful Auto Union Vows to Return and Win." AP. May 18. https://tinyurl.com/yy99ds3k

Leopold, Les. 2007. *The Man Who Hated Work and Loved Labor.* Chelsea Green.

Lichtenstein, Nelson. 2024. "The 'Stand-Up' Strike of 2023 Takes Its Place in UAW History." *New Labor Forum* 33, no. 2: 48–55. https://doi.org/10.1177/10957960241245944

Lichtenstein, Nelson. 2022. "Sectoral Bargaining in the United States: Historical Roots of a Twenty-First Century Renewal." In *The Cambridge Handbook of Labor and Democracy,* edited by Angela B. Cornell and Mark Barenberg. Cambridge University Press.

Logan, John. 2022. "High-Octane Organizing at Starbucks." *New Labor Forum* 31, no. 3: 36–42. https://doi.org/10.1177/10957960221117829

Logan, John. 2023. "A Model for Labor's Renewal? The Starbucks Campaign." *New Labor Forum* 32, no. 1: 87–94. https://doi.org/10.1177/10957960221144972

Mallon, Matthew J. 2024. "Embracing the Heat: 'Hot Labor Summer' Rekindles NLRB Authority, and the Need for Strategic Enforcement." *University of Miami Business Law Review* 33, no. 1. https://tinyurl.com/382hhm6u

Morena, Edouard, Dunja Krause, and Dimitris Stevis, eds. 2019. *Just Transitions: Social Justice in the Shift Towards a Low-Carbon World.* Pluto Press.

Morgenstern, Richard D., William A. Pizer, and Jhih-Shyang Shih. 2002. "Jobs Versus the Environment: An Industry-Level Perspective." *Journal of Environmental Economics and Management* 43, no. 3: 412–436. https://doi.org/10.1006/jeem.2001.1191

National Labor Relations Board. No date. "Election Reports, FY2014–2025." https://tinyurl.com/2xkr545c

National Labor Relations Board. 2023. "Tesla Inc." Case Number 03-CA-312449. Region 03, Buffalo, New York. https://www.nlrb.gov/case/03-CA-312449.

Naughton, Nora, and Grace Kay. "How Tesla's New Pay Levels Compare to Unionized Workers at Ford and GM." *Business Insider.* February 9. https://tinyurl.com/5bj852e9

Newell, Peter, and Dustin Mulvaney. 2013. "The Political Economy of the 'Just Transition.'" *The Geographical Journal* 179, no. 2: 132–410. https://doi.org/10.1111/geoj.12008

Occupational Safety and Health Administration. No date. "OSHA Establishment Search." https://tinyurl.com/yn7a6b2d

Räthzel, Nora, and David L. Uzzell. 2013. *Trade Unions in the Green Economy: Working for the Environment.* Routledge.

Raymond, Nate. 2024. "Massachusetts Voters Allow Uber, Lyft Drivers to Unionize." Reuters. November 6. https://tinyurl.com/hu56xhed

Satheesh, Silpa. 2020. "Moving Beyond Class: A Critical Review of Labor–Environmental Conflicts from the Global South." *Sociology Compass* 14, no. 7. https://doi.org/10.1111/soc4.12797

Subramanian, Pras. 2024. "Rivian Stock Drops as Full-Year Loss Projection Widens, But 'Modest Gross Profit' Still Expected in Q4." *Yahoo Finance*. November 8. https://tinyurl.com/9c9xcc9k

Townsend, Keith, Adrian Wilkinson, and John Burgess. 2013. "Is Enterprise Bargaining Still a Better Way of Working?" *Journal of Industrial Relations* 55, no. 1: 100–117. https://doi.org/10.1177/0022185612465533

United Auto Workers. 2024. "Joint Statement from UAW and Volkswagen on Certification of Election Results in Chattanooga–UAW | United Automobile, Aerospace and Agricultural Implement Workers of America." April 30. https://tinyurl.com/2avdk72f

United Auto Workers. No date. "Backgrounder on Big Three Bargaining." https://tinyurl.com/mrktss5c

United Auto Workers. 2023a. "High Risk & Low Pay, Hazardous Conditions and Low Wages Show Standards Must Be Raised at Battery Cell Plants Getting Billions in Taxpayer Dollars, A Case Study of Ultium Cells in Lordstown." November. https://tinyurl.com/yc2ebna8

United Auto Workers. 2023b. "UAW Wins Just Transition at General Motors, October 6. https://tinyurl.com/y48jh7wc

UAW Local 1112. 2024. "UAW–Ultium Cells Local Agreement." June. https://tinyurl.com/235duvv9

UAW Research. 2019. "Taking the High Road: Strategies for a Fair EV Future." March. https://tinyurl.com/4r92e553

U.S. Court of Appeals for the Fifth Circuit. 2023. Case No. 22-60493. November 14. https://tinyurl.com/y3tk22fr

U.S. Court of Appeals for the Fifth Circuit. 2024. Case 21-60285. October 25. https://tinyurl.com/ycf352zz

Vachon, Todd E. 2023. *Clean Air and Good Jobs: U.S. Labor and the Struggle for Climate Justice.* Temple University Press.

Weststar, Johanna, and Marie-Josée Legault. 2017. "Why Might a Videogame Developer Join a Union?" *Labor Studies Journal* 42, no. 4: 295–321. https://doi.org/10.1177/0160449X17731878

Yang, Zhijie, Haibo Huang, and Feng Lin. 2022. "Sustainable Electric Vehicle Batteries for a Sustainable World: Perspectives on Battery Cathodes, Environment, Supply Chain, Manufacturing, Life Cycle, and Policy." *Advanced Energy Materials* 12, no. 26. https://doi.org/10.1002/aenm.202200383

# Industrial Environmental Policy: Markets, Labor, and the Rhode Island Experiment

PATRICK CROWLEY

*Rhode Island AFL-CIO*
*Climate Jobs Rhode Island*

## Abstract

In this chapter I discuss the passage of the Rhode Island Act on Climate, an experiment in state-based industrial environmental policy. I argue that the approach taken by the State of Rhode Island, spurred on by groups like Climate Jobs Rhode Island, is a significant departure from neoliberal ideology without neatly fitting into current conversations around environmental policy. To demonstrate this, after providing a narrative of how the Act on Climate came to be enacted, I discuss how much of the debate about the direction of environmental policy in the United States fits into a counterproductive Marxist/capitalist binary. However, this chapter argues that the approach taken in Rhode Island is not some middle ground between these two poles but something altogether different. The Act on Climate legislated into existence state-based mandates for economy-wide decarbonization goals. The act, along with a package of legislation enacted over the course of the subsequent two years, puts the state into position to act as market creator. As the creator of a new energy market, the state is setting both the supply and demand curves of this new economy. If the intersection of those curves determines market prices, Rhode Island, by enacting strict labor standards in the marketplace, is determining that profits from renewable energy projects are to come from innovation and efficiency rather than exploitation and manipulation. As with all experiments, data will need to be collected, assessed, and debated to determine whether this approach to industrial environmental policy is ultimately successful, but, I argue, this heterodox approach to state-based market creation is a tool worth exploring.

## INTRODUCTION

A truism of American political dogma views the individual states of the union as laboratories of democracy. This concept, formulated by Supreme Court Justice Louis Brandeis, suggests that these experiments can be both "social and economic," and because they are confined to a single state, they are "without risk to the rest of the country" (*New State Ice Co. v. Liebmann* 1932). According to Brandeis, a single courageous state can, if its citizens choose, potentially show the rest of the states in the union how to create a more just nation. The State of Rhode Island is currently engaged in one of these experiments related to the climate crisis. If successful, it could demonstrate how to effectively address the threats posed by climate change through the careful crafting of industrial environmental policy centered on citizen engagement and state involvement in marketplace creation.

In what follows, I outline this experiment's nature and argue that it does not fit neatly into current climate strategy discussions. As I see the current conversation unfolding, there are two main threads in the discussion. First, there is the thread looking at the world around us and seeing our political leaders at best only tinkering with adaptation and mitigation schemes that are inadequate in the face of an existential crisis. The second thread is more optimistic about our chances to make meaningful strides on reversing the damage done by climate change at both the national and international levels, and posits that we are well on our way toward solving the climate crisis if we continue our current trajectory.

An interesting feature of this binary conversation about our current situation is that the two camps can more or less fit into a Marxist/capitalist ideological divide, with the Marxist (or at least Marxist-infused) thinking dominating the first group— believing we are not doing enough and therefore doomed—and the capitalist-orientated group occupying the more optimistic space. I argue that the Rhode Island experiment in industrial environmental policy is neither. It's not a middle ground landing somewhere between an imagined left/right economic spectrum but something different altogether, adhering more to a U.S. labor union tradition of being neither Marxist nor capitalist while acknowledging the strengths and weaknesses of both.

## THE RHODE ISLAND ACT ON CLIMATE

In the spring of 2021, Rhode Island Governor Daniel McKee signed into law the Rhode Island Act on Climate. The law requires the State of Rhode Island to develop "strategies, programs, and actions to meet economy-wide enforceable targets for greenhouse gas emissions reductions" of 45% below 1990 levels by 2030, 80% below 1990 levels by 2040, and net-zero emissions by 2050 (Title 42 State Affairs and Government, R.I. Gen. Laws §42-6.2-2). The legislation was supported by every major environmental organization in the state, as well as the Rhode Island AFL-CIO and the Rhode Island Building and Construction Trades Council (Crowley 2024: 74). When the law was passed, one of the lead architects of the legislation, Rhode Island state senator and then-chair of the Senate Committee on the Environment Dawn Euer, said "[t]he Act on Climate represents a strong commitment to the long-term health of our planet, as well as economic opportunity for our state. With this act, we are jumping to the leading edge of those states and nations that are changing the landscape of power generation" (State of Rhode Island 2021).

The Act on Climate does three critical things. First, it mandates that the State of Rhode Island, through a previously established Executive Climate Coordinating

Council populated by the heads of 13 key state agencies, develop a five-year plan to meet the net-zero emission standard by December 31, 2025, then produce a new plan every five years thereafter (State of Rhode Island 2024). Second, it reinvigorates a private citizen–led advisory board, as well as a science and technical advisory board, to complement and check the work of the government-based Coordinating Council. Third, it establishes what needs to be included in each five-year plan and who must participate in the formulation of the plan..

The five-year plans must, according to the law, demonstrate how the state intends to address the needs of communities disproportionately impacted by climate change, including

- How to address past inequities in environmental and public health policy
- How to provide support for workers impacted in the transition away from legacy fuel-related employment
- How the state will create family-sustaining clean energy jobs with wages and benefits meeting or exceeding existing area standards

The plan must also explain how the state intends to recruit women; Black, Brown, and Indigenous people; veterans; formerly incarcerated people; and people living with disabilities into the new clean energy jobs. Furthermore, the plan must also demonstrate how these impacted communities had a hand in influencing the plan and requires the state to produce regular public progress reports on the goals outlined in the plan (Title 42 State Affairs and Government, R.I. Gen. Laws §42-6.2-2).

Critically, the law is enforceable through the right of civil action brought against the State of Rhode Island by the office of the Rhode Island Attorney General and/or any resident, company, or organization, including labor unions (Title 42 State Affairs and Government, R.I. Gen. Laws §42-6.2-10). This provision is modeled on the federal Clean Water Action Act. As early as January 1, 2026, if even a single resident of the state believes the plans established by the state are not meeting the goals of the Act on Climate, they may file suit against the state in Rhode Island Superior Court. The court is also empowered to award successful litigants reasonable attorney fees and expert witness fees.

## LEGISLATIVE SCAFFOLDING

In the summer of 2022, only a year after the Act on Climate became law, the State of Rhode Island began legislating into existence the "strategies, programs, and actions" necessary to create an economy designed to decarbonize our environment. First, Governor McKee signed into law legislation requiring "one hundred percent … of Rhode Island's electricity demand [be] from renewable energy by 2033" (State of Rhode Island 2022a). State Representative Deborah Ruggiero, lead sponsor of the legislation, said, "This bill supports renewable energy growth and is consistent with the Act on Climate's goal of reducing carbon emissions by to [sic] net-zero by 2050. In addition to reducing emissions and our reliance on fossil fuels that must be brought to Rhode Island from other places, creating renewable energy supports the green industry, creating thousands of good paying jobs right here in Rhode Island."

Next, Governor McKee signed legislation requiring Rhode Island's private energy utility, Rhode Island Energy, to issue a procurement of between 600 and 1,000 megawatts of additional offshore wind supply (State of Rhode Island 2022b). The governor said at the time, "Adding offshore wind clean energy capacity is essential for meeting our new 100 percent renewable energy by 2033 goal and our Act on Climate emissions reductions target." When completed, the offshore wind created by this procurement would meet up to 30% of Rhode Island's energy demand, powering 340,000 homes. Another way to calculate the scope of this procurement: according to the U.S. Census Bureau (2023), there are 486,029 housing units in Rhode Island, so this offshore wind supply could potentially power 70% of the housing units in the state.

Last, the governor signed into law a bill sponsored by the president of the Rhode Island Senate and retired staff member for the Laborers International Union of North America in New England, Dominick Ruggerio, titled "Labor Standards in Renewable Energy Projects" (Title 39 Public Utilities and Carriers, R.I. Gen. Laws §39-26.9-1). This legislation mandates that whenever the state, or a subdivision of it including municipalities, is a market actor in a renewable energy project (that is, when a project receives a public financial incentive or any form of tax relief or tax subsidy), that project must adhere to three labor standards.

First, the project developer must enter into a labor peace agreement with any unions involved in "the construction, installation, use, maintenance, operation, changing, or retiring of a renewable energy resource" (Title 39 Public Utilities and Carriers, R.I. Gen. Laws §39-26.9-2). A labor peace agreement is a written agreement between an employer and a union saying the employer will not disrupt a union's attempt to organize workers in exchange for the union agreeing not to disrupt the employer's operation through strikes, pickets, or other means.

Second, the labor standards law requires prospective bidders on any covered project valued at $5 million or more to have a tradeworker apprenticeship program approved by the State of Rhode Island. Additionally, if the bidder is to employ five or more workers, the bidder must ensure that at least 15% of the work hours are performed by apprentices. To ensure that the bidder has a legitimate apprenticeship program, the law requires bidders to demonstrate that they have not previously been sanctioned by the state for defaulting on any project or misclassifying workers (Title 39 Public Utilities and Carriers, R.I. Gen. Laws §39-26.9-3).

The apprenticeship standard is vital to the equity-based goals of this new industrial environmental policy. As I have written elsewhere, the demographics of the construction workforce are changing (Crowley 2024). Male, White, and older (construction workers in Rhode Island are five years older than the median age of all workers) workers are being replaced not by their sons, but by Black, Brown, and female workers (National Association of Home Builders 2023). As the workforce diversifies while simultaneously transitioning from legacy fuel-based jobs to renewable energy jobs, it would not be a desired outcome (from a labor union point of view at least) if the family-sustaining wages of the unionized, largely White and male workforce did not accrue to the workers of the future just because of their gender, skin color, or nation of origin. By inserting the apprenticeship standard into the scaffolding of this new economy, Rhode Island is attempting to manage this transition from the worker up, ensuring that new workers get the opportunity to work through this economic transition without sacrificing the economic gains made by unionized workers who came before them. And because nearly all construction trade unions operate their own apprenticeship programs, it will be the labor movement that is intentionally diversifying this new economy.

Third, in addition to the labor peace agreement standard and the apprenticeship standard, the law requires potential bidders to pay prevailing wages in the construction, operations, and maintenance phases of the project. Prevailing wages, established at the federal level by the Davis–Bacon Act, are the combination of hourly wages, overtime, and benefits paid to a majority of workers in a particular geographic area such as a city, county, or state (Fact Sheet #66: The Davis–Bacon and Related Acts (DBRA) 2023). Rhode Island workers are also protected by strong, state-based prevailing wage laws as well and, given the high union density in the state, for most crafts and trades involved in renewable energy construction, it is the union wage rate that establishes the prevailing wage. Furthermore, the Rhode Island Department of Labor and Training requires employers to submit certified payroll records under penalty of perjury to ensure that workers are indeed getting paid fairly (Rhode Island Department of Labor and Training 2024).

When you consider the 100% renewable energy standard law, the offshore wind procurement law, and the labor standards law together, what you begin to see is the emergence of a new type of economy based on fundamentally different values than existing economic structures. Said another way, what the state of Rhode Island has done is legislate into existence the supply and demand curves for its net-zero emissions economy. By passing the 100% renewable energy standard law, the State of Rhode Island created the demand curve and set a deadline of 2033 for it to be fully developed. By passing the offshore wind procurement law, the State of Rhode Island created the supply curve. If the intersection of these curves establishes the process of commodity price discovery, which forms the basis of exchange between purchases and sellers, what the third piece of this new industrial environmental policy (the labor standards piece) does is establish that any profits derived from the price of commodities in this market (i.e., renewable energy) must be derived from innovation and efficiency instead of exploitation and manipulation. As Susan AnderBois, the director of climate and government relations for the Nature Conservancy in Rhode Island, testified during the debate on the labor standards law, "[W]e need to treat the Earth well, and we also need to treat the people of the Earth well" (Capital TV 2023).

Thus, the State of Rhode Island is radically recreating its economy by establishing a new industrial environmental policy aimed at driving change in the energy marketplace. What we have done, as a state, is look at what we need (an economy that does not contribute to the destruction of the planet through carbon-fueled climate change), democratically determined a course of action on how to get there (enacting the Act on Climate), then legislated the economic components we need to meet our goals (the 100% renewable energy standard, the 1000 MW offshore wind procurement, and labor standards). Learning from the experience of other economic experiments, we collectively determined this economy will not reward market actors who derive profit from paying workers the least possible wage to secure their labor. Rather, we determined workers should earn the highest wage and benefit level possible. By doing so, we are ensuring not only that working people are treated fairly but also that this emerging market will be guided by state law to be more innovative and efficient, which will drive energy costs down *and* meet the emission goals we have set out for ourselves as a state and as a people.

It is important to point out that good legislation alone will not ensure the success of the Act on Climate or the Rhode Island experiment. While the process of planning strategies, programs, and actions to meet the goals of the Act on Climate must include workers and other front-line communities, the state has a power advantage in terms of resources, including staff, time, and expertise, that most nongovernmental organizations do not possess. Likewise, relying on the citizen lawsuit protection alone to force the state to do the right thing, while important, is not a perfect check on its power. For a lawsuit to be successful, it will rely on a judge's determination—a judge who will bring their own biases and understanding of the issues to the discussion. Legal decisions can be years in the making, and the inevitable appeals will add time to the process of resolving any disputes—years that could further delay meeting the goals of the Act on Climate.

So, parallel to the Rhode Island legislative and market experiment is a role for advocacy by nongovernmental organizations (NGOs), including those working in the intersectional space of environmental activism and labor rights. One such organization is one I helped found in 2021, Climate Jobs Rhode Island (CJRI). CJRI was formed through the partnership of the Rhode Island AFL-CIO and several of the state's leading environmental organizations, including the Nature Conservancy Rhode Island, Green Energy Consumers Alliance, and the Rhode Island Audubon Society, to advocate for a just transition to a green economy. In only a few years, CJRI has changed the discussion about labor unions and environmentalists and worked to overcome media-generated differences and conflicts between the goals of each movement.

## THE STATE AS MARKET CREATOR

It is important to think through just how sweeping this new industrial environmental policy really is. The 2050 net-zero emission goals are not unprecedented: fourteen other states have enacted similar goals (Cohen 2023), as has the federal government under President Joe Biden (Council on Environmental Quality, no date). But imbued within this project is a radical departure from the prevailing economic thinking of the past 50 years. Under the economic philosophy generally referred to as "neoliberalism," governments intentionally withdrew from an active role in managing the economy.

There are many definitions of neoliberalism, but this simple one from economic journalist and professor Robert Kuttner is helpful:

> Neoliberalism's premise is that free markets can regulate themselves; that government is inherently incompetent, captive to special interests, and an intrusion on the efficiency of the market; that in distributive terms, market outcomes are basically deserved; and that redistribution creates perverse incentives by punishing the economy's winners and rewarding its losers. So government should get out of the market's way. (Kuttner 2019)

As a result, Japanese scholar Kohei Saito notes in his book *Slow Down: The Degrowth Manifesto* that "[n]eoliberalism promoted privatization, deregulation and austerity while expanding financial markets and free trade, setting the world on the path to globalization" (Saito 2024).

In contrast to neoliberalism, the State of Rhode Island, through the Act on Climate, intentionally seeks to direct the economy toward a specific goal—the elimination of carbon emissions from the Rhode Island economy. It requires the state not just to tinker with the existing energy marketplace but to establish "strategies, programs, and actions" to steer the economy toward the net-zero emission goal (Title 42 State Affairs and Government, RIGL §42-6.2-2 (2)(i)). It further allows individuals or organizations to appeal to the judicial branch of state government if they feel the executive or legislative branches of government are not adequately making the changes to the economy called for in the law. And the judicial branch of government is empowered to require the other branches to interfere more in the economy if they affirm a petitioner's claims.

In his 2011 book, *Debt: The First 5000 Years*, the late anthropologist David Graeber writes that "[s]tates created markets. Markets require states. Neither could continue without the other, at least, in anything like the forms we would recognize today" (Graeber 2011: 71). While not expressly discussed at the time, what Rhode Island is doing is turning away from the dominant neoliberal economic theories to assume a more heterodox interventionist orientation. Rhode Island, through this new industrial environmental policy is, *as a state*, creating the market mechanisms necessary to produce a desired economic outcome—namely, a net-zero emission economy. But more importantly, *as a people*, we have the tools to hold the state accountable for its economic behavior.

Rhode Island is also using this new industrial environmental policy to create an economy with specific rules of exchange. The French historian Fernand Braudel wrote in his book *Afterthoughts on Material Civilization and Capitalism* that the rules of exchange are "'transparent' exchanges, which involve no surprises, in which each party knows in advance the rules and the outcome, and for which the always moderate profits can be roughly calculated beforehand" (Braudel 1977: 50). By mandating labor standards, the State of Rhode Island is telling market actors in advance the rules for this economy must ensure that workers are treated fairly, that underrepresented populations must enjoy redress of past and ongoing grievances, that the state will enforce these rules to produce the desired outcomes, and if the state doesn't, the citizens can use the judicial branch of government to force the executive and legislative branches into compliance.

Not only is Rhode Island's experiment a break with neoliberalism's obsession with what it considers "free markets," it is also a course correction in the thinking that, from an economic point of view, the climate crisis is a series of ongoing either/or binary propositions. In the infancy of the climate change debate, the *Ur* either/or proposition was either climate change is real—or it is not. Subsequently, the either/or binary shifted to ask whether climate change is humanmade or not. Now that the consensus thinking in the debate has embraced the reality that climate change is real and caused by human activity, people now wrestle with ideas of what to do about it.

The "what to do about it" part naturally, but not exclusively, focuses on economics because it is chiefly human economic activity causing climate change. So, if human economic activity is the root cause of climate change, how then to change economic activity to stop climate change from progressing and eventually recede? Several recent works are important to consider, but as I will show, they also mostly adhere to a similarly stagnant either/or trajectory. In other words, the new either/or proposition is either capitalism is the reason we have climate change so we must move away from capitalism or it—capitalism—is the solution to climate change, and we must embrace it to save ourselves.

A reader might think with my argument's reliance on the language of markets and supply and demand that I fall in the latter camp. I do not, and I believe the Marxist/capitalist binary is counterproductive. Rhode Island's experiment in industrial environmental policy is something altogether different. It is not my goal in this chapter to claim that the new industrial environmental policy or the democratic enactment of market tools is simply a newer more egalitarian version of neoliberalism. On the contrary, I think it is a more dramatic departure from that ideology than even some of the advocates understood at the time. I'm not an economist—I'm a labor organizer trained as a historian—but when I look at the ongoing discussion in the literature of climate change, I see the either/or matrix as a barrier to progress and hope to see the Rhode Island experiment as a way through the blockage.

## CURRENT POPULAR DISCUSSIONS IN ENVIRONMENTAL ECONOMICS

For example, in the book *Slow Down: The Degrowth Manifesto* already referenced above, Saito, writing in the Marxist tradition, argues that Marx's thinking about the economy evolved dramatically over the course of his writings and that "conventional Marxists" who focus on the importance of continual economic growth miss the point (Saito 2024: 224). To Saito, Marx's later writing on subjects such as ecology argues that sustainable communities (or communes—it's not really clear whether the two words are interchangeable) are a form of resistance to capitalism (Saito 2024: 119–120). Therefore, by "[b]ringing about the democratization of production and the shortening of work hours, for example, must include the participation of labor unions" (which, as a union organizer, I am sympathetic with) can free Marxism from engaging in the same exploitative economic practices of the capitalist economies. (Saito 2024: 232).

But for me, a key point of disagreement with Saito is the role of what he refers to as "politicalism" plays in addressing climate change. Politicalism, according to Saito, is "the belief that if we simply select good leaders within a framework of representative democracy, we can leave it up to these politicians and experts to put optimal policies and laws in place for us" (Saito 2024: 132). Saito's politicalism argument is too reductivist, and Rhode Island's industrial environmental experiment explains why. To enact this law, yes, elected leaders in the Rhode Island legislature needed to engage in the regular legislative process of bill drafting, public hearings, floor debate, and law passage—but that only scratches the surface of what it took to create this new policy. Hundreds if not thousands of Rhode Islanders, along with dozens of NGOs including labor unions, environmental organizations, and community and social justice organizations, were critical to shaping what this new policy looked like, working collaboratively with elected officials to draft legislation, craft the policies, and push back against criticism and counterproposals from status quo–minded interests. Was it strikes, picket lines, and the seizure of factories? No, but I assert that the effort to enact the Act on Climate certainly can be interpreted as a form of class struggle.

Operating in the same Marxist tradition but inverting the theoretical approach is another recent work titled *Climate Change as Class War: Building Socialism on a Warming Planet* by Mathew Huber. For Huber, contra Saito, to properly address the climate crisis, it "has to mean more for the many and less for the few" (Huber 2022: 32). The "many" here being the vast majority of human beings on Earth—precariously employed workers and subsistence farmers. The "few" are not just the top one percent of income earners but the conventional bogeyman in this kind of argument—the owners of production. Huber makes an important point at this stage in his argument. He highlights a 2017 study documenting that as few as 100 companies have been responsible for 71% of all emissions since 1988 (Huber 2022: 23). To counter this concentration of climate change–producing economic power, Huber proposes a three-pronged approach: (1) the struggle against climate change needs to focus on industrial, not individual, production of carbon emissions, (2) this climate struggle needs to decenter the professional class and reprioritize the working class, and (3) the climate struggle needs to be grounded in a mass popular movement (Huber 2022: 3–6).

All three of these prongs are part of the Rhode Island experiment in industrial environmental policy. But where I diverge from Huber is in his understanding of market mechanisms and the path toward accomplishing the worker-centered decarbonization goal. Like Marx's own treatment of the issue of supply and demand in *Wage Labor and Capital*, Huber's analysis of the concepts, and by extension markets themselves, is lacking. In a smart but ultimately disappointing chapter about market-based attempts to impact the course of climate change, he writes, "[t]he key to market-based climate policy is this faith in the invisible hand of the marketplace" as an indirect force that can direct corporations and consumers alike to change their behavior (Huber 2022: 53). He goes on to say, "[f]or market exchange to work, and to be fair, it is the dispersed, atomized choices who must act without more power than any other individual" (Huber 2022: 57).

That is precisely what we are experimenting with here in Rhode Island— democratically predetermined rules of the game where the economic and political power is intentionally distributed, or as Huber says, dispersed. Huber correctly identifies that, given the current national and global weakness of the "left," seizing the means of production is not a real possibility. But requiring states and economies to emphasize the "public good over private profit" is a worthy step forward (Huber 2022: 291). And, while he is critical of "technocratic" attempts to influence decarbonization efforts such as clean electricity standards, what we are doing in Rhode Island with our renewable energy standard is going one step farther by attaching pro-worker conditions to those standards.

So, while they may operate in the same Marxist tradition, Huber and Saito map entirely different paths forward. But in addition to sharing a tradition, they also share a sense of urgency that if the people of the Earth continue on the path we are headed, then in a time not too distant into the future the Earth with burn and we will all die. Or not all of us: the rich will do fine, but the poor and the working class will be obliterated.

I note this common theme in Saito and Huber because if we look at work operating outside of and against the Marxist tradition, we find a shared optimism about what humans are currently doing to counteract climate change thanks to capitalism. A pair of books (both, incidentally, featuring front cover endorsements by Microsoft CEO Bill Gates), *Climate Capitalism: Winning the Race to Zero Emissions and Solving the Crisis of Our Age* by Akshat Rathi, and *Not the End of the World: How We Can Be the First Generation to Build a Sustainable Planet* by Hannah Ritchie, offer much more optimistic interpretations of what humans are doing to address the negative impacts of climate change and thus are able to avoid the call for sweeping paradigm shifts like Saito and Huber ask us to consider.

In *Climate Capitalism*, Rathi writes, "[t]his book shows why it is important to harness the forces of capitalism to tackle the climate problem— and how that work has already begun" (Rathi 2023: 3). Over a series of 12 chapters, Rathi, an organic chemist by training but a journalist by trade, interviews various experts on the international climate stage who each, according to the book, are demonstrating that climate change is fundamentally a "market failure." The chapters are interestingly deployed to highlight the three areas in which Rathi believes capitalism can solve the climate crisis: technology, policy, and people (Rathi 2024: 10).

The problem is Rathi never really says what this "capitalism" that is going to save the world actually is. He acknowledges early on that capitalism as we currently experience it is broken and that fixing it "will mean not just reforming how business is done but completely transforming some industries" (Rathi 2024: 10). Later in the book, he includes a few paragraphs on this theme by criticizing one of the fathers of neoliberalism, the economist Milton Friedman, even going so far as to characterize the climate crisis in part spawned by Friedman's work as a "corruption of capitalism" (Rathi 2024: 200). But he is very clear that solutions like Saito and Huber propose are weak cases that have disastrous historical antecedents.

Instead of degrowth or socializing the economy, Rathi argues that "capitalism, from the United States to China, in now deeply entrenched." Therefore, the goal should be to reform this existing system to meet the Earth's climate goals (Rathi 2024: 180). Ironically, this is a conclusion not dissimilar to Huber's. Both Rathi and Huber acknowledge that the proletariat seizing the means of production is not happening anytime soon, so there must be some other way to re-engineer an economy away from rapacious profit first/climate second–driven outcomes. As Rathi writes in his conclusion, "[g]etting to net zero emissions on a deadline will mean changing everything" (Rathi 2024: 203).

But what is the best way to "change everything?" Is it to create a sense of crisis and danger in a target audience so people will be compelled to act? Or does change happen when an audience is hopeful and optimistic about the possibility of a brighter and better future? Rathi doesn't answer, but Hannah Ritchie, a global development researcher at Oxford University, believes it is the latter. In her book, *Not The End of the End of the World: How We Can Be the First Generation to Build a Sustainable Planet*, Ritchie argues that the "last generation" mindset— that is, the idea that we are the last generation to live on this planet unaltered by climate change, is a psychological barrier to making progress on confronting climate change. Furthermore, "last generation" thinking not only leads to a sense of paralysis among people looking at the impact of climate change on humans, but it is also bad science (Ritchie 2024: 12).

Delightfully iconoclastic, *Not the End of the End of the World* is, like Huber and Rathi, decidedly anti-degrowth. "The [math] doesn't check out," she writes, arguing that there are too many poor people outnumbering rich ones to achieve a fairer standard of living "through redistribution alone" (Ritchie 2024: 34). She also argues that new technology will allow us to create a sustainable future where "[e]conomic growth is not incompatible with reducing our environmental impact" (Ritchie 2024: 36). But while Ritchie does conclude that finding a sustainable balance between growth and a livable planet requires that "we need to change political and economic incentives," her arguments that incentives and technology alone are enough to produce the carbon-free society are in the end unconvincing, in part because of the arguments raised by another English writer, Brett Christophers.

In his 2024 book, *The Price Is Wrong: Why Capitalism Won't Save the Planet*, Christophers, a geographer specializing in the impact of capitalism on human beings, argues that we cannot expect private industry to continue to drive the decarbonization process because there is not enough money to be made for investors. For years, he argues, we have been waiting for the time when the cost of producing renewable energy would be cheaper than the cost of producing carbon-based energy. "Renewables crossed that apparent Rubicon in the late 2010s," but, he argues, we are still not making adequate progress towards our renewable energy production goals. Why not? "The hurdle [is] profitability." While it might be cheaper to produce renewable energy through sources such as offshore wind and solar, it is still not generating enough profits to compel private energy firms to make the switch (Christophers 2024: ix–xiii).

A key contribution to the economics of climate change debate in *The Price Is Wrong* is the way Christophers describes the energy marketplace. Like Graeber and Braudel, Christophers believes that markets "are often actively fashioned by government hands, and this is certainly true of electricity" (Christophers 2024: 37). This means the marketplace isn't some other-worldly site of exchange of goods for currency but rather an "administrative construct." These complex systems might govern the generation and distribution of energy, but they also dictate the parameters of profit for the generators and distributors. So, despite "all of those supply and demand curves in economic textbooks," relying on the market alone to drive the change toward decarbonization will not produce the desired outcome (Christophers 2024: 100).

While I read Christophers to be sympathetic to ideas in line with Huber about government ownership of energy companies in order to fully meet national climate goals, he makes a compelling case for "a potential third way of sorts" (Christophers 2024: 371). Instead of direct state ownership, since energy markets are already a creation of the state, the state could "compel" the market players to move quicker on the shift to renewable energy generation and distribution. Sadly, Christophers too quickly dismisses this idea as too heterodox to be effective. But this is exactly the heterodoxy that is built into our Rhode Island experiment, and we will see if it is indeed possible.

I think Christophers and the other writers featured in this chapter could benefit from a key insight that David Graeber offers us in his reading of Fernand Braudel. Yes, Braudel does write that "[c]apitalism only triumphs when it becomes identified with the state, when it is the state" (Braudel 1977: 64), but Graeber, correctly in my view, interprets Braudel to mean that markets and capitalism are not the same thing and in fact "could equally well be conceived as opposites" (Graeber 2011: 260). In this view, the inherent conflicts in a market economy are not between the state and capitalism but between the free market and capitalism. So, a prerequisite of a free market is a strong state to serve as a countervailing force against the monopolistic intentions of capitalists.

The cultural hegemony imposed by neoliberal economic ideology has diverted attention away from this concept for decades. It is not my intention here to completely revisit the economic history of the past 50 years, never mind the past 150 years, but I believe we inherently used to know the point Braudel and Graeber make. It is one reason progressive public policy in the United States was once centered on anti-monopoly regulation and trustbusting—not to enhance the power of the state against free markets but to protect free markets from those who would weaken them by dominating them. Also, it is not the role of this chapter to offer a complete analysis of the value of Braudel and Graeber's economic viewpoints. But, like Huber and Christophers acknowledge, for better or worse we endure the conditions of a capitalist economy; so, if we want to change it or at least make it work for different ends, we need to understand what it is and what it is not.

## CONCLUSION

In the spring of 2024, two new reports seem to indicate that we are starting to see progress in attempts to push back against an economy prioritizing profit ahead of survival. According to the clean energy think tank Ember, 30% of the world's energy production was from renewable sources in 2023 (Altieri 2023). Simultaneously,

THE RHODE ISLAND EXPERIMENT

carbon-powered energy demand fell 12% from 2007, when Altieri's report argues it hits its peak. Also, a recent report by Synapse Energy Economics, commissioned by the Sierra Club, argues that in New England, where I live, if we can produce 9 gigawatts of offshore wind by 2030, the average utility customer will see their electricity bill drop a few dollars a month, saving, on a regional basis, up to $630 million in electrical costs (Whited, Knight, Kwok, and Sylva 2024). This is encouraging, but if we are truly going to prioritize human and planetary well-being ahead of shareholder profit, governments must act deliberately to realign our economic priorities.

What I argue above is that the State of Rhode Island, pushed by its citizens, is engaged in a social and economic experiment in industrial environmental policy aimed at just such a realignment. To meet its goals of establishing a carbon-neutral state economy by 2050, Rhode Island is using the power of state government to create an energy marketplace, allowing it to develop a new economy not reliant on power produced from fossil fuel but from renewable energy sources. Through intentional policy decisions, the State of Rhode Island is erecting the scaffolding necessary for market mechanisms such as supply and demand to work with an agreed-upon set of rules, known to market agents in advance, and directed toward a desired set of outcomes. Like any experiment, data will be produced and analyzed in order to report any conclusions relating to the success or failure of the experiment.

The state has also created the mechanisms to guide the experiment and afforded itself the ability to make necessary adjustments if targets and goals are not met. More forcefully, if the people of Rhode Island believe the state is not conducting the experiment as they like, they can use the power inherent in their democratic rights to affect the changes they desire.

A key to understanding and potentially replicating this experiment is an appreciation of the economic mechanisms at work—namely, the concepts of supply and demand. To reach the proper understanding, a certain familiarity with the current concepts impacting the climate change discussion, especially those concerned with human economic activity, is required and supplied above. This chapter doesn't fit squarely within the current conversation regarding what to do about climate change, and it offers, through an understanding of the Rhode Island experiment in industrial environmental policy, a different view. This view, that the state can create the marketplace for energy it wants to through democratic means, is heterodox to conventional economic theories but nevertheless coming to life in Rhode Island. Therefore, if this experiment is successful, it will demonstrate a new way to reach our climate goals.

## REFERENCES

Altieri, Katye. 2023. "Tracking National Ambition Towards a Global Tripling of Renewables." Ember Energy. November 21. https://tinyurl.com/3d9bur2n

Braudel, Fernand. 1977. *Afterthoughts on Material Civilization and Capitalism*. Johns Hopkins University Press.

Capital TV. 2023. "Testimony of Susan AnderBois." March.

Christophers, Brett. 2024. *The Price Is Wrong: Why Capitalism Won't Save The Planet*. Verso Press.

Cohen, Rona. 2023. "States with Net-Zero Carbon Emissions Targets." The Council of State Governments, Eastern Regional Conference. March 23. https://tinyurl.com/yckxn8pt

Council on Environmental Quality. No date. "Net-Zero Emissions Operations by 2050, Including a 65% Reduction by 2030." Federal Sustainability Plan. https://tinyurl.com/yraytnd4

Crowley, Patrick. 2024. "Organizing Climate Jobs Rhode Island." In *Power Lines: Building a Labor–Climate Justice Movement*, edited by Patrick Crowley and Lindsay Zafir. The New Press.

"Fact Sheet #66: The Davis-Bacon and Related Acts (DBRA)." 2023. WHD Fact Sheets. October. https://www.dol.gov/agencies/whd/fact-sheets/66-dbra

Graeber, David. 2011. *Debt: The First 5000 Years*. Melville House.

Huber, Matthew T. 2022. *Climate Change as Class War: Building Socialism on a Warming Planet*. Verso.

Kuttner, Robert. 2019. "Neoliberalism: Political Success, Economic Failure." *The American Prospect*. June 25. https://tinyurl.com/2fxyucem

Marx, Karl. 1847. "By What Is the Price of a Commodity Determined?" *Wage Labour and Capital*. https://tinyurl.com/3zbamvp5

National Association of Home Builders. 2023. "Is the Construction Workforce Older than Other Industries?" June 6. https://tinyurl.com/yyy9z375

*New State Ice Co. v. Liebmann*, 285 U.S. 262 (1932). Justia Law. https://tinyurl.com/2tds38up

Rathi, Akshat. 2024. *Climate Capitalism: Winning the Race to Zero Emissions and Solving the Crisis of Our Age*. Greystone Books.

Rhode Island Department of Labor and Training. 2024. "Prevailing Wage." October 18. https://tinyurl.com/tyrafcv5

Ritchie, Hannah. 2024. *Not the End of the World: How We Can Be the First Generation to Build a Sustainable Planet*. Little, Brown Spark.

Saito, Kohei. 2024. *Slow Down: The Degrowth Manifesto*. Astra House.

State of Rhode Island. 2021. "Press Releases Governor McKee Signs Act on Climate." Last modified April 14, 2021. https://tinyurl.com/yh4wjhaj

State of Rhode Island. 2022a. "Governor McKee Signs Historic Legislation Requiring 100% of Rhode Island's Electricity to be Offset by Renewable Energy by 2033." https://tinyurl.com/y3ufb4a6

State of Rhode Island. 2022b. "Governor McKee Signs Legislation Requiring Offshore Wind Procurement for 600 to 1,000 Megawatts." https://tinyurl.com/43ccj3z7

State of Rhode Island. 2024. "EC4 Overview." https://tinyurl.com/5ew44n65

"Title 39 Public Utilities and Carriers." Labor Standards in Renewable Energy Projects R.I. Gen. Laws § 39-26.9-1. https://tinyurl.com/ynnfh3mw

"Title 39 Public Utilities and Carriers." Labor Standards in Renewable Energy Projects R.I. Gen. Laws § 39-26.9-2. https://tinyurl.com/9b9ttskc

"Title 39 Public Utilities and Carriers." Labor Standards in Renewable Energy Projects R.I. Gen. Laws § 39-26.9-3. https://tinyurl.com/ye25t584

"Title 42 State Affairs and Government." R.I. Gen. Laws § 42-6.2-2. https://tinyurl.com/3upattvp

"Title 42 State Affairs and Government." R.I. Gen. Laws § 42-6.2-10. https://tinyurl.com/4u2hmtv5

U.S. Census Bureau. "QuickFacts Rhode Island." 2023. https://tinyurl.com/3d9s6cxw

Whited, Melissa, Pat Knight, Shelley Kwok, and Patricio Sylva. 2024. "Charting the Wind: Quantifying the Ratepayer, Climate, and Public Health Benefits of Offshore Wind in New England." June 3. https://tinyurl.com/3redetnm

CHAPTER 9

# Building a Diverse, Equitable, and Unionized Clean Energy Workforce: Best Practices and Lessons Learned

ZACH CUNNINGHAM
MELISSA SHETLER
*Cornell ILR Climate Jobs Institute*

## Abstract

As climate-fueled natural disasters continue to increase in frequency and effect, working people and front-line, environmental justice communities are often hit first and worst. To address these issues, we must transition to a clean energy economy at the pace and scale science demands, protect existing workers, and create opportunities for new workers to access good paying, community-sustaining careers. Apprenticeship readiness programs are a key component to making this happen. Apprenticeship readiness advances diversity in the construction trades by supporting women and historically marginalized communities to access union career pathways. Successful programs partner with community organizations to recruit participants, provide them with training and wraparound services as they learn about different trades, and offer support as participants apply for apprenticeship and matriculate into the unionized building trades. And as construction in climate and clean energy continues to grow, apprenticeship readiness also helps provide a pipeline of skilled workers needed to make the energy transition successful. This chapter uses lessons learned from three case studies to highlight the key takeaways for developing and implementing successful workforce development interventions. Apprenticeship readiness and registered apprenticeship programs—like those in the building and construction trades—are a scalable and replicable model for addressing workforce needs, growing the clean energy economy, and driving equitable creation of union jobs. Government, industry, and labor all have a role to play in driving investment to these successful models.

## INTRODUCTION

In November 2023, the authors of this chapter published a report titled "Building an Equitable, Diverse, and Unionized Clean Energy Economy: What We Can Learn

167

from Apprenticeship Readiness." In this report, we highlighted how the political and economic climate at the time "[presented] a rare opportunity to tackle the dual crises of climate change and inequality. … Through … the Infrastructure Investment and Jobs Act (IIJA) and the Inflation Reduction Act (IRA)," we wrote, "the federal government has allocated trillions of dollars for clean energy development and other infrastructure projects. Government leaders have also centered union job creation and equity. … Through initiatives like Justice40, the Biden administration has also committed itself to allocating significant benefits from federal legislation to traditionally underserved communities."

To put it lightly, times have changed. With Donald Trump retaking the White House, the federal government has deprioritized job quality standards and equity initiatives, such as Justice40. Leaders have pivoted strongly away from climate action and clean energy development, instead focusing on expanded fossil fuel extraction. It is yet to be seen exactly how badly clean energy and adjacent industries will be hurt, but it is safe to say that the most meaningful progress on the climate jobs front is likely to take place at the state and local levels over the coming years.

But some things remain constant. We are still facing a climate crisis that demands action from leaders at all levels of government and civil society. We are also still facing a crisis of inequality, limiting opportunities based on income, wealth, race, gender, and other areas of difference. And there is still significant support for growing high-quality climate jobs from state and local leaders, organized labor, activist organizations, market actors, and others. Now more than ever, proponents are looking for proven models to create a diverse, equitable, and unionized clean energy economy.

In this chapter, we will highlight one such model: apprenticeship readiness programs. We will draw extensively from our 2023 report to define apprenticeship readiness and highlight its role in providing the skilled workforce needed to build a robust construction sector and clean energy economy. We will also highlight some best practices gleaned from field research and outline recommendations for key stakeholders—including union leaders, employers, community leaders, policymakers, and philanthropic institutions—on how to expand successful apprenticeship readiness models.

Finally, we will situate our research in the current moment. How does a change in federal leadership alter the apprenticeship readiness landscape? How can this model expand to other critical sectors, including manufacturing and utilities? These are some of the questions we will discuss in this chapter.

## APPRENTICESHIP READINESS AS A CLIMATE JOBS INTERVENTION

### The Construction Industry, Apprenticeship, and the Centrality of Organized Labor

If we are serious about taking on climate change, there are few sectors more strategically placed than construction. Our built environment—buildings, energy systems, water infrastructure, transportation—is both made and *re*made by workers. Any economic transition, by necessity, will go through the construction industry. The unionized construction industry is different from other economic sectors in many ways. Construction firms tend to employ a small number of full-time tradespeople, while unions take on a significant role in both training and dispatching labor to job sites. It is common for tradespeople to have several different employers over the course of a given year, along with several periods of unemployment. When a construction project ends, workers will often return to the union hiring hall, where a designated person (or people) dispatches these workers to new workplaces, helping employers ensure a ready and skilled workforce is available to them when needed.

In this sector, the joint labor–management training center is the primary mechanism for ensuring there are sufficient skilled tradespeople to meet the demands of industry employers. These registered apprenticeship programs—or RAPs—are jointly funded and administered by organized labor and signatory contractors. They are registered with either the state or federal government, depending on location, and are heavily regulated. Training centers offer ongoing education and certification for journey-level workers who have completed their initial apprenticeship training, but the bulk of programming at these centers is geared toward training apprentices as they learn their chosen crafts.

Apprenticeship programs aim to match their recruitment schedules with industry demand. As such, many training centers do not offer a steady, consistent number of apprenticeship slots from year to year. In fact, during an economic downturn, apprenticeship programs can go extended periods of time without welcoming significant numbers of new people into union membership. Elected union leaders are invested in limiting the amount of time any given member spends out of work, and this focus helps drive decisions about apprenticeship recruitment schedules.

RAPs and unionized construction careers have created pathways to dignified, high-quality, well-paid work for millions of people over the years. A union card, especially in construction, has long been an important tool for working-class people to secure good jobs without a college degree. Furthermore, union apprenticeship programs typically come at no direct cost to participants, and those enrolled can "earn while they learn"—getting paid to complete hands-on construction work at jobsites during their apprenticeship. Wages, benefits, training, and safety standards on unionized construction sites far exceed those on nonunion sites.

Beyond improving wages and working conditions for union members, organized labor is also central to healthy operations in the construction industry. Unions play this role in several ways:

- Workforce development: According to North America's Building Trades Unions (NABTU), "[W]orkforce training is at the foundation of the core mission and existence of each building trades union." Labor runs over 1,900 apprenticeship centers throughout North America, investing close to $2 billion to train apprentices and journey-level workers. Roughly 71% of all construction apprentices go through joint labor–management training programs, which register around 73,000 new apprentices each year ("Apprenticeship & Training" 2024). Journey-level union members also use training centers to update certifications, receive training on new technologies, and gain new skills. According to a 2016 study from the Pew Research Center, 87% of workers believe it will be essential for them to get training and develop new job skills throughout their work lives in order to keep up with changes in the workplace ("The State of American Jobs" 2016). Organized labor provides much of the training infrastructure needed to make this happen.

• Dispatching skilled labor: Construction is an infamously fickle industry, with projects and labor demand fluctuating all the time because of changes in economic circumstances. Contractors typically do not have a large permanent workforce; instead, they rely on organized labor to control the supply of labor and dispatch workers through their hiring halls. This provides employers with flexibility, knowing they're likely to have the number of workers needed—no more and no less—at any given time.

• Industry stabilization: Through collective bargaining agreements and project labor agreements, unions set industry standards for pay, benefits, and working conditions. This provides a level of certainty to union contractors, locking in fixed costs and allowing for long-term planning. By establishing standards across trades, unions help prevent a race to the bottom, where contractors compete to win projects by lowering costs (often through cuts to wages and benefits) rather than delivering efficient, quality work.

## Exclusionary Practices

The construction trades, nonunion and union alike, also have a history of exclusion and closing off opportunities to women, people of color, and other marginalized groups. Both employers and trade unions have contributed to this dynamic. As Figueroa, Grabelsky, and Lamare (2013) note:

> Dating back to the 1960s, the unionized construction industry was a focal point for the civil rights movement as communities of color witnessed a construction boom offering the false promise of good jobs for urban residents. Because of discriminatory hiring practices, the overwhelming majority of union construction jobs went to white workers. In New York City, for example, 92 percent of building trades union members [were] white. Some of the skilled trades had virtually no African American members.

In addition to the discriminatory hiring practices that can serve as barriers to entry in the trades, informal practices also unintentionally serve to maintain homogeneity. Traditionally, word of mouth has been a major element of recruitment into trade unions. People's networks tend to be populated with others who look like them or come from similar backgrounds. As people spread the word about career opportunities to their family members and friends, union membership can skew toward certain demographics.

In recent years, there have been some major examples of progress toward diversifying the trades. In 2017, the Economic Policy Institute found that in New York City, the "union construction sector employs a greater share of Black workers and pays them more than the nonunion construction sector, and unions are drawing many more Blacks into construction through apprenticeships compared to 20 years ago" (Mishel 2017). But still, women and people of color make up a small percentage of unionized building trades members nationwide. White workers make up 75% of union membership in construction (Feliciano Reyes 2022), while the number of women nationally in the construction trades remains in the single digits (Hegewisch and Mefferd 2021).

## Apprenticeship Readiness

So, what is apprenticeship readiness, and what role does it play in expanding access to unionized construction careers? The U.S. Department of Labor defines it as "a program or set of strategies designed to prepare individuals to enter and succeed in a Registered Apprenticeship program." A more robust definition comes from PowerSwitch Action (formerly the Partnership for Working Families) in the *Construction Careers Handbook* (Partnership for Working Families 2013):

> **Pre-apprenticeship programs** recruit and orient new workers, help them identify the apprenticeship program most suited to them, prepare them to take the test [*author's note: apprenticeships do not universally require testing, but it is a common practice*] and support their initial career efforts. In addition to orientation to the industry, they sometimes provide other kinds of support including life skills training, financial literacy and job readiness. Some pre-apprenticeship or pre-training programs provide stipends to help pay for tools and equipment, and may even offer help with transportation and childcare.

Many places, such as PowerSwitch Action, use the term "pre-apprenticeship" rather than apprenticeship readiness, and different organizations and areas of the United States have their own preferences about which term to use. In this report, we will use the terms interchangeably, most-often leaning toward "apprenticeship readiness."

Since the primary goal of apprenticeship readiness is to prepare people to enter and succeed in an apprenticeship, their structures typically do not mirror traditional workforce development programs. To familiarize participants with the industry, programming reflects the realities of working in construction. Program leaders and their industry partners frequently connect in order to understand new market trends and workforce needs. Open lines of communication help ensure that curricula and recruitment strategies mirror changing circumstances on the ground.

Apprenticeship readiness programs tend to share several features. The Aspen Institute (Conway and Gerber 2009; Conway, Gerber, and Helmer 2010) identified several key curriculum elements, including

- Basic construction skills training
- An introduction to one or more trades
- Construction math, tool identification, and blueprint reading
- Test preparation services
- Workplace safety
- Visits to apprenticeship training centers and active construction sites
- Job readiness training and "soft skills"
- Various certifications, including OSHA 10, flagging, and CPR
- Case management services to help participants navigate the apprenticeship application process
- Mentorship and connections with tradespeople from similar backgrounds
- Financial services such as stipends and travel vouchers

In addition to classroom and hands-on training, programs typically also provide a range of support and career exploration services. Support services range from stipends, travel vouchers, and other financial support to more hands-on help such as case management or mock interviews. Participants visit worksites and engage with industry leaders to identify the right career path for them.

Apprenticeship readiness programs vary in their structures, target populations, and curricula. For example, some programs simulate construction workers' typical work hours, while others offer more flexible training on a part-time basis, at night or on weekends. Many programs use NABTU's Multi-Craft Core Curriculum (MC3) as a standard, but even this program allows for some flexibility and optionality.

The quality of apprenticeship readiness programs varies, but high-quality programs have led to several positive outcomes for participants, such as

- Increasing participants' knowledge of the construction industry and its career opportunities
- Providing participants with technical skills and hands-on experience in the construction trades
- Helping participants develop soft skills such as communication, teamwork, and problem solving
- Improving participants' employability and job readiness
- Increasing participants' likelihood of entering and succeeding in apprenticeship programs
- Achieving higher career earnings and union career opportunities

Program leaders and staff can recruit in any manner and with any target audience they choose. However, apprenticeship readiness has become one of the primary industry tools for diversifying the construction trades in recent years. Labor unions, employers, and other partners have invested time and energy into setting up programs that target any number of underrepresented communities. In fact, NABTU explicitly states that their MC3 program is "a gateway for local residents—focusing on women, people of color, and transitioning veterans—to gain access to Building Trades' registered apprenticeship programs" ("Apprenticeship Readiness Programs" 2023). Finally, several apprenticeship readiness programs have so-called direct entry status with the U.S. Department of Labor and in partnership with local building trades unions. This means that successful graduates of designated apprenticeship readiness programs are able to fast-track in some capacity and enter directly into an apprenticeship should they meet that trade's requirements. This is an important tool in diversifying the trades because it allows apprenticeship readiness graduates from diverse backgrounds a more direct path into apprenticeship and construction careers. However, many apprenticeship readiness programs have not established the agreements with building trades unions and government bodies that are required to implement direct entry.

Apprenticeship readiness is one of the most valuable tools available for recruiting and orienting underrepresented populations to the building and construction trades. When done well—and when combined with other efforts to aid apprenticeship retention and graduation, job placement, career advancement, and a sense of belonging on the jobsite—it can help expand access to high-quality construction careers.

## Building the Clean Energy Economy

The vast majority of apprenticeship readiness programs, including the ones we examined in our report, do not have an exclusive focus on clean energy. The MC3 curriculum has incorporated elements of green construction, but most programs have a more general focus on helping participants navigate the construction industry writ large.

This broader orientation makes sense because apprenticeship programs teach people a craft not a job. Apprenticeship intentionally gives participants a wide range of transferable skills that they can use to build a long-term career. Journalist Lee Harris (2022) wrote about this dynamic in the solar industry, observing that

> Training a "solar installer" rather than an electrician is like training a "burger-flipper" rather than a chef. It confers a limited number of skills, which are only appropriate for the current industry. And it does not correspond to any Department of Labor–certified apprenticeship program, so it could be a dead end for workers who could otherwise flip to electrical work in another sector.

However, there is good reason to believe that construction workers will be seeing increasing work hours on projects tied to the clean energy economy in the years ahead. In the 2023 report, "Equity in Focus: Job Creation for a Just Society," authors Anne Marie Brady, Risa Lieberwitz, and Zach Cunningham (2023) outline the extent of work that will need to take place:

> We need to massively scale-up our energy efficiency, retrofit efforts and decarbonize buildings. We need to expand public transportation, electrify most of our transportation fleet, and

figure out how to transition long-haul and heavy transport to low-carbon fuels. The amount of electricity we generate from solar, wind, nuclear, and other low-to-zero carbon energy sources is nowhere near sufficient. And our electrical grid needs to be three to five times larger than it currently is to accommodate electrification in the transportation and buildings sectors. We are not creating a brand new economy from scratch, but we do need a strategy to transform our economy and grow clean energy to the scale science demands of us.

Any effective strategy to tackle climate change will require massive job creation, especially in the construction sector. Over 8 million people worked in an "energy job" in 2023, according to the U.S. Department of Energy. Energy jobs include "employees of a qualifying firm that spend some portion of their time supporting the qualifying building, construction, and/or retrofitting of energy infrastructure projects." The Department of Energy recognizes five categories of energy jobs: electric power generation; energy efficiency; fuels; motor vehicles; and transmission, distribution, and storage. Nearly half of all job growth in the energy sector last year came in clean energy (Jones, Zamora-Duran, and Lipman 2024).

Recent efforts at the federal and state levels have accelerated these trends. At the federal level, the Infrastructure Investment and Jobs Act (IIJA) of 2021 and the Inflation Reduction Act (IRA) of 2022 have poured trillions of dollars into clean energy development, decarbonization, adaptation and resilience, and other infrastructure priorities. While the recent passage of the "One Big Beautiful Bill Act" (OBBB) under the Trump administration dealt a blow to many of the programs supporting the buildout of clean energy projects, such as solar and wind, support remains for the expansion of geothermal and thermal energy networks, carbon capture projects, nuclear, and green hydrogen. Additionally, many states and cities are continuing to focus on meeting their clean energy goals.

For example, many northeastern states have agreed to procure significant amounts of electricity produced by renewable sources, and Illinois passed its Climate and Equitable Jobs Act (CEJA) in 2021, which commits millions to solar installations, workforce development, and other areas. A 2020 study by Rewiring America estimates that an aggressive electrification and decarbonization strategy could lead to 25 million jobs created over 15 years (Griffith, Calisch, and Laskey 2020).

With ongoing discussions in the media and academia about a workforce shortage, many wonder where exactly these workers will come from. Joint labor–management training centers are a key part of the puzzle. Existing training centers, with some tweaks, are equipped to teach many of the skills needed for clean energy and energy efficiency work. To complete this work at scale, though, they will need to significantly increase the number of apprentices and journeyworkers in the coming years. Apprenticeship readiness is an important tool in making this happen.

## BEST PRACTICES AND RECOMMENDATIONS FOR SCALING EFFECTIVE APPRENTICESHIP READINESS PROGRAMS

It takes a village to build a successful apprenticeship readiness program. Organized labor, construction employers, and community organizations all have central roles in shaping and successfully operating a program. Other parties, such as government agencies and philanthropic institutions, also provide resources and help build capacity so programs can thrive.

In our original field research, we examined three apprenticeship readiness programs with unique geographies, structures, and target audiences. Our research included interviews with program staff, union and industry leaders, policy makers, program participants, community organizations, and others intimately familiar with the funding and day-to-day practices of each organization. It also involved visits to the construction sites, apprenticeship centers, and classrooms where participants experienced programming. Table 1 (next page) presents details about the three programs we studied in depth.

We also supplemented this fieldwork with desk research and interviews with many other apprenticeship readiness leaders and advocates across the United States. In studying these programs, the authors had three goals in mind:

- Illuminating "best practices" for practitioners who are interested in starting or expanding their work in the apprenticeship readiness space
- Providing a series of recommendations for three key stakeholders—organized labor, construction employers, and community-based organizations—on how best to approach apprenticeship readiness work in order to maximize impact
- Offering further recommendations for governmental policy makers and private funders, both of whom can fund apprenticeship readiness and build organizational capacity

## BEST PRACTICES TO OPERATE SUCCESSFUL APPRENTICESHIP READINESS PROGRAMS

### Industry-Driven Programming, Tied to Demand

The unionized construction industry is unique in many respects. It is common for people to have several different employers over the course of a given year, and firms tend to employ a small number of full-time tradespeople. Unions take on a significant role in both training and placing workers onto jobsites to make sure contractors have the workers needed to complete a project. Additionally, apprenticeship programs aim to match their recruitment schedules with industry demand, meaning many training centers do not offer a consistent number of apprenticeship slots from year to year.

Because of this, apprenticeship readiness cannot simply mirror traditional workforce development programs in structure or programming. Programming should be aligned with the realities of the construction industry, and it should also make sure participants deeply understand how the industry operates. This involves frequent communication between program leadership and industry partners—and staff members dedicated

Table 1. Overview and Summary of Three Apprenticeship
Readiness Centers Examined in This Study

| HIRE360 Chicago, Illinois | CityBuild San Francisco, California | Apprenticeship Readiness Collective, New York City |
|---|---|---|
| Established in 2019 | Established in 2005 | An umbrella organization for four unique programs: The Edward J. Malloy Initiative for Construction Skills, Nontraditional Employment for Women, Pathways to Apprenticeship, and New York Helmets to Hardhats |
| Nonprofit that targets recruitment to low-income communities of color, primarily on the south and west sides | City agency that recruits from low-income and high-minority ZIP codes throughout San Francisco | |
| Has "four pillars" that work in tandem: workforce development, building a diverse contractor base, building a diverse supplier base, and developing future builders through youth engagement | Rooted in the history of community activism around diverse hiring on construction projects | |
| | Has four key stakeholders: Community organizations, industry employers, labor unions and training centers, and city agencies | Each organization has its own unique target audiences and funding models, though there are many similarities in structure and curricula |
| Deep buy-in and partnerships with labor unions, training centers, employers, and community organizations | Helps San Francisco meet legally mandated local hiring requirements on public construction projects | Each program has developed strong relationships with community organizations and public agencies central to advancing opportunity for their unique audiences |
| Growing rapidly and expanding its footprint throughout Illinois | Long-established relationships in the industry | |
| Does not have direct entry status | Does have direct entry status | Each program has direct entry status |

to this role—to understand market trends and workforce needs. Programming, certifications, and recruitment strategies can then shift accordingly.

## The Right People at the Table

Successful programs center their key goal—expanding opportunities to high-quality construction careers—and work relentlessly to recruit the parties that *need* to be involved in order to drive success. This typically includes construction employers, building and construction trades unions, and some constellation of government agencies and community-based organizations. Each of these organizations plays a key role in driving equity in the industry:

- Government agencies can provide funding, and they can also set diversity mandates or goals that drive demand for apprenticeship readiness graduates.
- Community organizations often have strong relationships on the ground in underserved communities, giving them expertise on where to recruit participants and how to bring them into programs.

- Unions and apprenticeship programs accept new apprentices and invest in training them on their chosen crafts.
- Employers hire both apprentices and journey-level workers to carry out construction projects.

Collaboration is not always easy. There is often deep-seated skepticism or outright hostility among parties based on past experiences. But many organizations we examined saw important breakthroughs after undergoing a clear-eyed assessment of the key players in the local construction industry—and then working to bring these parties to the decision-making table.

## Shared Infrastructure Among Partners

Once the right parties are at the table, they should share resources to maximize impact. These resources can be *physical*, like tools and classroom space. For example, CityBuild leases classrooms and uses instructors from the City College of San Francisco. Sharing *human capital* is also critical. It is very common for unions and employers to donate staff time to teach courses, lead mock interviews, and take other measures to develop participants' knowledge and skills. *Relational* resources are also key, such as providing programs access to community contacts that can expand opportunities for participants.

Apprenticeship readiness programs also help build capacity within their partner organizations. CityBuild and the City of San Francisco provide extensive funding to community organizations to carry out their recruitment and case management efforts. In 2024, HIRE360 opened a new 45,000 ft² warehouse, office, and training building where they provide space for unions to conduct solar installation training and for small-scale suppliers to store inventory as they expand their operations.

It is also common for successful apprenticeship readiness programs to partner with one another to maximize impact. New York City's Apprenticeship Readiness Collective formed as an umbrella for multiple programs to share resources, lessons, and best practices from their work with different populations in the field. HIRE360 has also partnered extensively with Chicago Women in the Trades (CWIT) and other organizations. One HIRE360 staffer noted the mutual benefits of this relationship, as CWIT has strong bonds with community organizations while HIRE360 has built deeper relationships with industry partners.

## Clarity About Populations Served

With finite resources available, programs cannot cast as wide of a net as they may like. Therefore, identifying a key population to serve and focusing resources on doing that work effectively is typically a winning strategy.

HIRE360 focuses its recruitment efforts on Chicago's south and west sides, where the city's high-poverty neighborhoods and communities of color are concentrated. Each program in the Apprenticeship Readiness Collective has a primary (if not *exclusive*) focus—high school students, women, veterans, and returning citizens.

CityBuild largely focuses its recruitment in high-poverty, high-unemployment ZIP codes in the city of San Francisco. The program's community partners are also place based, largely serving populations in specific neighborhoods in need of high-quality career opportunities.

## Credible Messengers

As mentioned previously, word of mouth has traditionally been a major element of recruitment into the unionized building and construction trades. While this practice has its benefits, it can also (oftentimes unintentionally) exacerbate homogeneity, as people's social networks are often less diverse than the population as a whole.

Apprenticeship readiness can turn this paradigm on its head. The most effective evangelists for programs are oftentimes their own graduates. Building a roster of program graduates and people from underrepresented backgrounds willing to share their experiences and spread the word about career opportunities can help drive recruitment and retention.

## Clear Metrics of Success and Data Tracking

Each program examined has a key set of metrics to gauge effectiveness. The most common metrics include program graduation rates and apprenticeship or job placement rates. And given the programs' emphasis on diversity and equity in the construction industry, they also tend to track demographic information of program participants, most notably race and gender.

However, leaders should avoid an *overreliance* on data that can lead to inflexibility when adapting to changing circumstances in the field. Program leaders note that not every person who drops out of a program represents a "failure." A major goal of apprenticeship readiness is to familiarize participants with the industry, which means that some people will realize that construction is not the right career fit for them. The most successful programs often find a reasonable balance between setting measurable goals, implementing a robust data tracking operation, and knowing when it is appropriate to look beyond data to analyze their effectiveness.

## Funding That Best Serves Participants' Needs

First, it is impossible to understate how important financial services are for successful apprenticeship readiness programs. It is incredibly difficult for low-income individuals to commit to several weeks of training and site visits without some form of financial aid.

But the most appropriate use of aid differs from place to place and audience to audience. Monthly stipends may not be as critical for high school students in a construction skills program as they are for a 30-year-old woman with two children going through the Nontraditional Employment for Women Program. Whether they provide stipends, travel vouchers, daycare support, or other measures, successful programs prioritize learning participants' particular needs and shaping financial assistance to best meet those needs.

## Sustainable Funding Models

Successful programs provide highly personalized services to participants. This requires significant amounts of money and staff time. There is not a one-size-fits-all approach to fundraising. HIRE360, for example, prioritizes diversifying its funding sources, aiming for a roughly even split among public agencies, private charity, and industry partners. CityBuild takes a different approach, relying on a public budget line item—not public or private *grants*—for all its funding. Regardless of funding strategies, successful programs prioritize securing sustainable funds to run their organizations for the long term.

## RECOMMENDATIONS FOR KEY PARTIES

### Organized Labor

*Establish Direct Entry Relationships with Apprenticeship Readiness Programs*

It is often difficult for aspiring apprentices without prior connections to navigate the application process and timelines for different trades. By entering into a "direct entry" agreement with an apprenticeship readiness program, unions can allow qualified graduates a more direct path into the trade of their choice, bypassing the oftentimes lengthy waiting lists for entry.

There is a misconception that by entering into a direct entry relationship with an apprenticeship readiness program, labor will be required to take *any* graduate. This is not the case. Apprenticeship programs still determine their own eligibility requirements and the number of slots they will open. Direct entry simply means that qualified applicants who come from an apprenticeship readiness program jump to the front of the line for admission. In our research, several people referenced apprenticeship coordinators whose views on direct entry changed after matriculating several high-quality applicants into their programs.

*Work with Apprenticeship Readiness Organizations to Better Track Apprenticeship and Career Outcomes*

Data tracking is important for apprenticeship readiness programs, but it often stops after an apprenticeship placement. Organized labor can help track longer-run retention and career-advancement statistics, giving all parties richer data by which to judge their effectiveness at diversifying the industry and advancing long-term career opportunities.

*Incorporate Apprenticeship Readiness Into an Overall Organizing Strategy*

Apprenticeship readiness is not a panacea. The construction industry is a complicated ecosystem with many different parts that must move together. To have maximum impact, apprenticeship readiness should be part of a larger organizing strategy that includes multiple avenues for bringing in new members *and* building trust with existing members.

Organizing happens on many levels, most notably sending staff or existing union members to nonunion workplaces and convincing contractors to sign on to collective bargaining agreements. But unions can also organize *existing* members, educating them about why apprenticeship readiness and direct entry are in their best interests. Expanding and diversifying membership can expand solidarity, bring new ideas into the labor movement, and better position union contractors to win bids from government and other project owners.

## Construction Employers: Create Employment Pathways for Graduates

Apprenticeship readiness is effective only if there are apprenticeship and job placement opportunities for graduates. Direct entry allows organized labor to open space for apprenticeship readiness graduates, but employers must also prioritize creating job opportunities for those graduates by instituting local hiring commitments, funding workforce development programs, and setting strong diversity standards through project labor agreements, collective bargaining agreements, and other avenues.

## Community Organizations: Enter New Partnerships

Community organizations play an important but very different role from organized labor and construction employers in the apprenticeship readiness ecosystem. While labor and employers control certain levers of entry, community partners tend to play more of a facilitation role, helping to recruit and provide services to people in the communities they serve as they navigate employment and training opportunities in the construction industry.

Community organizations often have complicated relationships with employers and organized labor. But if an apprenticeship readiness program meaningfully includes community organizations in decision-making structures, values their input and expertise, and shows a real commitment to diversifying the construction industry, participation from community organizations can be transformative for the industry and community alike. Keeping an open mind when engaging with new partners can make all the difference.

## Organized Labor, Construction Employers, and Community Organizations: Prioritize Providing Time and Resources

This recommendation cuts across all three types of organizations outlined above. While it may seem obvious, dedicating time and resources toward something many view as outside the scope of everyday work can be difficult. But one could argue that apprenticeship readiness is *essential* to long-term success of all parties and worthy of substantial investment.

## Policy Makers

*Invest in Growing Key Industries Like Clean Energy*

Apprenticeship readiness programs can succeed only if there is a robust construction market. A thriving market guarantees labor demand, opening more apprenticeship slots, and driving employers to bid on jobs and hire more workers. Policy makers can help to maximize the effectiveness of apprenticeship readiness programs by investing in key industries, creating a pipeline for graduates into registered apprenticeship programs and onto jobsites. Some areas where policy makers can make targeted investments are

- Making direct public investments in building clean energy sources
- Installing solar panels or other clean energy sources on government buildings and other publicly owned land
- Purchasing energy produced from wind, solar, and other clean sources from private companies
- Retrofitting and making upgrades to public buildings and transportation systems (including schools) to maximize energy efficiency
- Providing financial incentives for individuals and businesses to make energy efficiency upgrades to their buildings and vehicles
- Requiring energy efficiency and carbon-free energy sources on new construction
- Investing in expanding and upgrading public transportation systems
- Taking measures to make infrastructure more resilient to heat waves, extreme weather events, and other effects of climate change
- Repairing and replacing water lines and other crucial public infrastructure
- Hardening existing infrastructure to better withstand severe weather events and other effects of climate change

### Include Strong Labor and Equity Standards on All Publicly Funded Projects

Investing in key industries can help drive markets, but it will not necessarily create high-quality jobs. Whenever governments provide money or incentives for project development, there is an opportunity for policy makers to attach labor and equity requirements as a condition for receiving funds. Potential labor and equity standards include

- Prevailing wage and project labor agreement requirements for construction work
- Labor peace agreements on operations and maintenance work attached to publicly funded projects
- Wage floors for all work in a given area
- Community workforce agreements with targeted hiring goals
- Apprenticeship utilization requirements
- Local hiring and other targeted hiring policies

### Require "Labor Voice" on Workforce Development and Other Public Boards

Having advocates for high-quality job creation and equitable access in public office is critical to creating a diverse, equitable, clean energy economy and construction industry, but nothing replaces the experiences and perspectives of workers themselves. To make sure workers' interests are represented at all levels of government, policy makers can require "labor voice"—or a representative from organized labor—on key decision-making boards covering topics such as workforce development, energy and energy efficiency, construction, and economic development. Policy makers can also establish "just transition" boards and require voices from both organized labor and community organizations.

### Provide Adequate, Flexible Funding for Programs

Robust public funding allows programs to expand their footprint without sacrificing quality. But money should not come with too many strings attached. Successful programs allocate resources based on their knowledge of specific communities and what is needed for participants to succeed. Policy makers should follow the lead of apprenticeship readiness practitioners and avoid cumbersome requirements that ultimately detract from professionals in the field doing what they do best: serving participants and graduates.

## Private Funders

### Make Longer-Term Commitments to Fund Work

The transitory nature of foundation and philanthropic funding is a major source of stress for nonprofit leaders. In a field like apprenticeship readiness, where it takes time to build the relationships and identify the players in an industry, year-to-year funding can be a major impediment to program building. Private funders can experiment with longer-term funding commitments, freeing workforce development practitioners to analyze the industry, build relationships, and develop programs to make an impact over the long term.

The Arbor Brothers Grant Initiative based in the New York tri-state area, provides a good example. Arbor Brothers provides up to $300,000 of funding over three years ("Our Approach" 2024). They also invest 200 to 300 hours in consultancy for each grantee over 9 to 12 months. Consulting covers several areas key to organizational growth, such as executive coaching, strategic planning, financial management, and outcome measurements.

### Institute Flexible Metrics for Success

Similar to public funders, philanthropies can follow the lead of practitioners on the ground when determining the appropriate use of funds and metrics for success. It is important that programs set measurable goals, track progress, and use data to inform practice. But it is also important to balance this with reasonable exceptions on the basis of on-the-ground realities.

In one location studied, an apprenticeship readiness program helped place a participant in a union apprenticeship before the cohort graduated. They did so because the candidate was ready for the opportunity, the apprenticeship was in the participant's desired trade, and they did not know when the union would be taking applications again. However, leaders had to convince a funder that this was a "success" over the course of many conversations because the participant did not formally graduate from the program, which was one of the key metrics used to measure effectiveness.

In that same location, staff took an apprenticeship readiness class to a high-rise construction site early in the program. After taking participants to the top floor—which was an open-air construction site with no windows or walls yet installed—one person realized she was terribly afraid of heights. Even though she believed that a career in construction was right for her, the direct experience on a site led her to reconsider and leave the program. While she did not graduate, program staff viewed this as a success of sorts. After all, a major purpose of apprenticeship readiness is to familiarize participants with the industry and help them decide if it is right for them. It was much better for the participant to realize construction was not a good fit at that moment than after being accepted to an apprenticeship program and taking a spot that would be better used on someone else.

Apprenticeship readiness is meant to familiarize participants with the industry and place them into the career of their choice. This does not always happen on the expected timeline. And sometimes, people will drop out of a program after realizing it is not a good fit. These examples may count against graduation rates, but in some ways they are also a sign of an effective program.

*Help Programs To Build Capacity in Key Areas*
Many upstart nonprofits, including apprenticeship readiness programs, are filled with passionate advocates and subject-matter experts. However, these nonprofit leaders sometimes struggle to develop both the visionary practices *and* the day-to-day logistics needed to build long-term success. While it is important to fund service provision, investing in long-term capacity building—such as visioning, strategic planning, coaching, managing personnel, project management, and data tracking—is also key.

## THE FUTURE OF APPRENTICESHIP READINESS: THREATS AND OPPORTUNITIES

Even in the best of times, it is difficult to build effective apprenticeship readiness programs. Success requires a significant amount of work and coordination among different parties with their own interests and pressures, who often find themselves at loggerheads over key policy and economic development questions. With the Trump administration now in power, it is fair to wonder what the future holds for apprenticeship readiness.

Despite some challenges likely to come, there is still reason to believe that effective apprenticeship readiness programs can grow in the coming years. In the following pages, we will discuss some of the challenges on the horizon—and the opportunities to cement apprenticeship readiness as a key tool in building a diverse, equitable, and resilient climate and clean energy economy.

### The Role of State and Local Governments
Federal policies that prioritize climate action, equity, and high-quality job creation have come under attack in the Trump administration. A fossil fuel executive now leads the Department of Energy, and Trump himself has attacked renewable energy technologies such as wind turbines (Friedman and Plumer 2025). Further, conservative

pushback on diversity, equity, and inclusion initiatives has already had a chilling effect on businesses and other market actors (Norris 2024). While some labor leaders initially praised Donald Trump's choice of Lori Chavez-DeRemer to lead the Labor Department, his appointments and actions during his first term – such as stripping collective bargaining rights from scores of federal employees – show that he will not be a champion of union labor.

Even with an unfriendly federal administration, state and local governments can still advance an equitable and just transition. Where allowable, governments can set requirements for market participants looking to use public funds. These requirements can include several measures discussed previously, like prevailing wages, targeted hiring goals, community workforce agreements, and labor peace agreements. Governments can establish apprenticeship advisory boards that include labor, community, and industry experts whose oversight can ensure that the quality of programs is maintained, and that they continue to serve as bridges for historically disadvantaged communities into high-quality careers.

## Industry-Recognized Apprenticeship Programs
In 2017, then–President Trump signed Executive Order 13801, which sought to expand industry-recognized apprenticeship programs, or IRAPs. Registered apprenticeship programs favored by labor unions have significant government oversight. IRAPs were meant to shift oversight from the Department of Labor to industry partners, which critics allege would undermine safety and quality standards. President Biden overturned this rule in 2021 ("Fact Sheet: Biden Administration to Take Steps to Bolster Registered Apprenticeships" 2021).

Overturning Executive Order 13801 was not the only measure the Biden administration took to strengthen registered apprenticeships. For example, the Inflation Reduction Act (IRA) provided significantly larger tax breaks on clean energy projects for companies that employ a certain number of registered apprentices on their jobsites. The dual threat of reintroducing IRAPs and overturning portions of the IRA could seriously threaten the future of high-quality apprenticeship programs. And if the federal government undermines apprenticeship standards, apprenticeship readiness will also take a hit.

For these reasons, apprenticeship advisory boards and similar bodies outlined above will be especially important to institutionalizing labor voices and maintaining quality training around the country. With organized labor's support and guidance, state governments can also move to establish their own state-regulated apprenticeship systems as a safeguard against federal erosion of standards.

## Manufacturing
This chapter primarily focuses on the role of apprenticeship readiness programs that provide pathways into building and construction trades unions. This should still be a major focus moving forward, given construction's centrality to the energy transition and its long-established apprenticeship systems. However, manufacturing is also

critical to creating an equitable clean energy industry, and growth in this sector is essential to unlocking the job creation potential in a climate-safe economy. Growing manufacturing employment could also dovetail with the Trump administration's emphasis on growing industry in the United States.

According to a study by the International Energy Agency, global investments in clean energy technology and manufacturing grew by more than 70% from 2022 to 2024, with the United States and Europe showing the strongest investments and growth in battery manufacturing ("Advancing Clean Technology Manufacturing" 2024). Growth in the production of wind turbine components, photovoltaic solar panels and racking systems, electric vehicles and component parts, heat pumps, clean hydrogen electrolyzers, batteries, and other technologies provides the opportunity for an industrial build-out that could create long-term career opportunities for generations of workers.

With this in mind, several unions have invested heavily in strengthening apprenticeship training as an on-ramp to manufacturing careers. The United Auto Workers (UAW) represent manufacturing workers across the country in the automobile and other industries. Through its political action, organizing efforts, and collective bargaining agreements, the union has recently prioritized creating opportunities for unionized careers as car makers move toward producing electric and other low-emissions vehicles.

In 2023, the UAW founded the Center for Manufacturing a Green Economy (CMGE) with a $2 million-dollar federal award to bolster their efforts ("Department of Energy Awards $2 Million to UAW–CMGE" 2024). CMGE's first project is developing a workforce training program for a Sparkz lithium ferrophosphate battery manufacturing facility in Sacramento, California. According to CMGE's website, this program—made possible by a labor peace agreement between CMGE and Sparkz—"will build a robust and diverse recruitment pipeline from Sacramento's underserved communities, foundational job skills training, wraparound services to support cohorts, and a skills-based apprenticeship for production workers" ("CMGE Programs" 2024).

The International Association of Machinists and Aerospace Workers (IAM), another industrial union, represents 600,000 active and retired members in several industries. By establishing the Machinists Institute in 2019, IAM District 751 in Washington state has been a leader in growing registered apprenticeships in manufacturing. The institute has five registered programs across different industries, where apprentices receive 2,000 hours annually of on-the-job training, as well as 114 hours of supplementary training ("Apprenticeship Programs" 2025).

Growing apprenticeship programs in manufacturing also provide an opportunity to expand apprenticeship readiness in this space. The loss of manufacturing jobs in recent decades disproportionately hurt Black workers, with noncollege-educated Black manufacturing workers suffering the sharpest wage premium decline of any demographic group between 1980 and 2010 (Armstrong, Wilson, and Lowe 2024). The Machinists Institute shows that recruiting from diverse populations can help reverse this trend. In its "pre-employment training," its apprenticeship readiness equivalent, 18% of pre-apprentices in 2024 were women, while 73% were people of color ("Machinists Institute Impact Report" 2024).

## Funding

In addition to passing legislation with clear emissions benchmarks and climate goals, policy makers can also invest in the workforce training infrastructure needed to make these goals a reality. They can develop frameworks to ensure that investments target programs with a track record of placement into long-term, high-quality careers, which prioritize supportive wraparound services for front-line communities and communities of color. They can also invest in apprenticeship readiness by passing standards that establish minimum hours, standardized curricula, have articulated agreements with registered apprenticeship programs, and are prioritizing the recruitment of front-line communities and underrepresented workers.

State programs like Washington's cap-and-invest programs generate revenue by limiting overall carbon emissions, requiring businesses to acquire allowances for their emissions, and allowing businesses to purchase additional allowances from those who do not use their allotment ("Washington's Cap-and-Invest Program" 2025). Billions of dollars could be generated from these programs, and labor standards and workforce development requirements attached to this money can help drive an equitable transition.

Illinois' CEJA provides another model worth considering for policy makers. Passed in 2021, CEJA is a comprehensive energy policy that accelerates and incentivizes the buildout of renewable energy and electric vehicle charging infrastructure, while simultaneously investing in a statewide workforce development system and financial support meant to allow front-line communities and existing energy workers to more easily participate. The program's barrier reduction fund, for example, provides funding based on what an individual needs in order to complete training or maintain new employment, including transportation, childcare, tools, phone and utility bills, or other areas of need ("Climate and Equitable Jobs Act" 2025).

With the federal pullback on climate, workforce, and equity funding, private philanthropy will take on a new level of importance for apprenticeship readiness programs. There have been promising developments on this front. The Workers and Families Fund, for example, recently established a new Powering Climate and Infrastructure Careers for All Fund, which aims to create at least one million new jobs in the transition to a clean energy economy. The Workers and Families Fund provides support for two distinct areas: A training and career pathway, and a government planning and implementation pathway ("Powering Climate & Infrastructure Careers for All" 2025).

Philanthropic funding can also allow apprenticeship readiness programs to develop new curricula linked to climate action and new technologies. J.P. Morgan Chase recently provided funding for the Climate Jobs National Resource Center and the Climate Jobs Institute to work with partners in New York City's Apprenticeship

Readiness Collective to develop a clean energy curriculum for programs to integrate into their existing training.

## Childcare Support

Quality, affordable, and accessible childcare is one of the biggest barriers to more diverse populations entering careers in the trades and manufacturing. Childcare is critical to working families, yet the cost is often prohibitive and access limited. Both the construction and the manufacturing industry present additional challenges. Split shifts in manufacturing can mean overnight or long hours, and construction workers are typically expected to be on site by 6:30 A.M., with last-minute overtime requests common as projects push to meet deadlines. Childcare access for workers outside of the traditional 9 to 5, five-day work week is particularly challenging. One study found that only 8% of the center-based providers surveyed reported that they offer childcare during nonstandard hours (Dobbins et al. 2019).

California's Equal Representation in Construction Apprenticeship (ERiCA) was developed to help address some of these issues. A grant program housed within the state's Department of Industrial Relations, ERiCA recognizes the challenges in balancing work and childcare, dedicating funds toward childcare support for pre-apprenticeship and apprenticeship programs registered with California's Division of Apprenticeship Standards. According to ERiCA's website, funds "can be used to support stipends intended to pay for childcare, or to cover the cost of childcare coordination or in-house childcare for the families of the participants." The program also awards funds to organizations that prioritize community outreach, ensuring underrepresented populations are well-connected in the construction ecosystem ("Equal Representation in Construction Apprenticeship (ERiCA) Grant" 2025). Other states can replicate this model—and even expand on it to include manufacturing and other key sectors.

The CHIPS and Science Act, which was signed into law in 2022, provides another example. This landmark federal industrial policy focuses on bolstering semiconductor manufacturing and innovation in the United States. The CHIPS Act requires companies that receive direct federal funds of more than $150 million to build or expand semiconductor manufacturing facilities provide access to affordable, high-quality childcare for both the construction and long-term workers ("CHIPS Child Care Requirement" 2023). This requirement was developed with an understanding of the significant role that childcare plays in enabling workforce participation, particularly for mothers. Like other federal legislation passed during the Biden administration, though, the future of CHIPS is uncertain.

Figure 1. Apprenticeships Haven Grown Across Industries

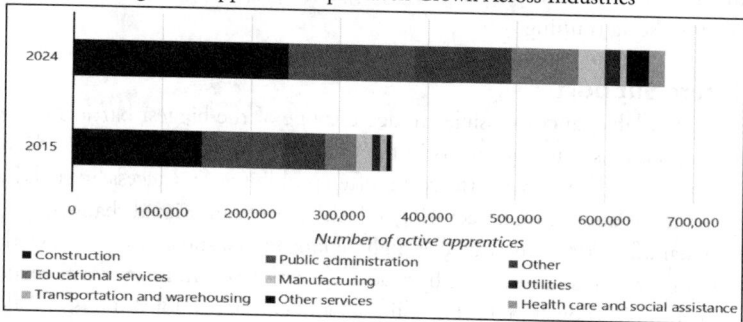

Council of Economic Advisers. Sources: RAPIDS; CEA calculations as of October 23, 2024.
Note: Public administration not covered in the economic census. "Other services" does not
include publica administration. "Other" includes observations without reported identity.

## CONCLUSION

In January 2025, climate scientists detailed some disconcerting, though not necessarily
surprising, findings. The year 2024 was the hottest in recorded history. Worse yet,
average global temperatures surpassed 1.5°C above pre-industrial levels for the first
time, a critical threshold under the 2015 Paris Agreement (Gramling 2025).

As if we needed more reminders of this moment's urgency, at the time of this
writing, climate-fueled wildfires are ripping through the city of Los Angeles, showing
that even our most iconic metropolises are not beyond the reach of disaster.

With the climate crisis worsening each day, policy makers and industry leaders
must take bold action to decarbonize the economy, build new sources of clean energy,
and harden our infrastructure to the effects of extreme weather. How we respond as
a society in the next few years will determine the health of our planet for generations.

But climate action should not happen on the backs of working people and front-line
communities. Instead, leaders must help create millions of high-quality, unionized
jobs—and expand access to these jobs—for populations typically left behind. In fact,
one could argue that doing so is a political imperative, given election results in 2024 that
swept many opponents of climate action into power. Building a clean energy economy
with low-road, precarious jobs will only exacerbate inequality and undermine support
for decarbonization in communities throughout the United States.

The unionized construction trades have been a source of family- and community-
sustaining jobs for decades. Through joint labor–management training centers,
organized labor has helped produce generations of highly skilled tradespeople to
complete important construction projects. While it is essential to acknowledge
discriminatory practices—both past and present—within the construction industry,
apprenticeship readiness programs provide a viable path forward for diversifying the
building and construction trades and supplying the labor force needed to perform
climate work into the future.

The clean energy economy is already here, and market forces indicate that it is
likely to grow in the coming years regardless of who is in power at the federal level.

Apprenticeship readiness can be one part of a larger organizing strategy to expand clean energy work; enlarge union market share; and attract, train, place, and retain a growing number of high-quality workers from diverse backgrounds.

## REFERENCES

"Advancing Clean Technology Manufacturing: An Energy Technology Perspectives Special Report." 2024. International Energy Agency. https://tinyurl.com/bdeazm8m

"Apprenticeship Programs." 2025. Machinists Institute (blog). https://tinyurl.com/3dmn9sfp

"Apprenticeship Readiness Programs." 2023. North America's Building Trades Unions. https://tinyurl.com/2sffpa5y

"Apprenticeship & Training." North America's Building Trades Unions. 2024. https://tinyurl.com/3v6prdsj

Armstrong, B., M.D. Wilson, and N.J. Lowe. 2024. "Race and the Manufacturing Workforce: Opportunities to Expand Growth and Equity in a Rebounding Industry Sector." W.E. Upjohn Institute for Employment Research. https://tinyurl.com/3s8fswdv

Brady, A.M., R. Lieberwitz, and Z. Cunningham. 2023. "Equity in Focus: Job Creation for a Just Society." ILR Worker Institute. https://tinyurl.com/ys4bx3te

"CHIPS Child Care Requirement: How Equitable Implementation Can Promote Stable, Well-Compensated Early Childhood Jobs." 2023. Center for the Study of Child Care Employment. https://tinyurl.com/ku97p6px

"Climate and Equitable Jobs Act." 2025. Illinois Department of Commerce & Economic Opportunity (blog). https://tinyurl.com/pncwdann

"CMGE Programs." 2024. UAW Center for Manufacturing a Green Economy (blog). https://tinyurl.com/yaabcp57

Conway, M., and A. Gerber. 2009. "Construction Pre-Apprenticeship Programs: Results from a National Survey." The Aspen Institute. https://tinyurl.com/bdd2knp7

Conway, M., A. Gerber, and M. Helmer. 2010. "Construction Pre-Apprenticeship Programs: Interviews with Field Leaders." The Aspen Institute. https://tinyurl.com/y87x2cks

"Department of Energy Awards $2 Million to UAW-CMGE to Build High Road Training Pathways for America's Climate Workforce." 2024. UAW (blog). https://tinyurl.com/yftzkjhr

Dobbins, D., K. Lange, C. Gardey, J. Bump, and J. Stewart. 2019. "It's About Time: Parents Who Work Nonstandard Hours Face Child Care Challenges." ChildCare Aware of America. https://tinyurl.com/yxk5ssj7

"Equal Representation in Construction Apprenticeship (ERiCA) Grant." 2025. State of California Department of Industrial Relations (blog). 2025. https://tinyurl.com/2py4788s

"Fact Sheet: Biden Administration to Take Steps to Bolster Registered Apprenticeships." 2021. The White House. https://tinyurl.com/2zanaxjy

Feliciano Reyes, J. 2022. "Broken Rung." *The Philadelphia Inquirer.* August 4. https://tinyurl.com/4tbe3hd2

Figueroa, M., Grabelsky, J., and R. Lamare. 2013. "Community Workforce Agreements: A Tool to Grow the Union Market and to Expand Access to Lifetime Careers in the Unionized Building Trades." *Labor Studies Journal* 38, no. 1: 7–31. http://dx.doi.org/10.1177/0160449X13490408

Friedman, L., and B. Plumer. 2025. "Trump Promises to End New Wind Farms." *New York Times.* January 7. https://tinyurl.com/274nx29e

Gramling, C. 2025. "2024 Was Earth's Hottest Year on Record, Passing a Dangerous Warming Threshold." *Science News.* January 10. https://tinyurl.com/3bexm8uh

Griffith, S., S. Calisch, and A. Laskey. 2020. "Mobilizing for a Zero Carbon America: Jobs, Jobs, Jobs, and More Jobs." *Rewiring America*. July 29. https://tinyurl.com/2s4xjjtm

Harris, L. 2022. "Workers on Solar's Frontlines." *American Prospect*. December 7. https://tinyurl.com/yhwkz9pm

Hegewisch, A., and E. Mefferd. 2021. "A Future Worth Building; What Tradeswomen Say about the Change They Need in the Construction Industry." Institute for Women's Policy Research. November. https://tinyurl.com/3dbw33wc

Jones, B., A. Zamora-Duran, and Z. Lipman. 2024. "2024 U.S. Energy & Employment Report." U.S Department of Energy. October. https://tinyurl.com/v2ebc7pw

"Machinists Institute Impact Report 2024." 2024. IAMAW. https://tinyurl.com/um6aw3mu

Mishel, L. 2017. "Diversity in the New York City Union and Nonunion Construction Sectors." Economic Policy Institute. March 2. https://tinyurl.com/ej2ctxxh

Norris, C. 2024. "How Some Companies Are Scaling Back DEI Initiatives after Conservative Backlash." PBS. August 27. https://tinyurl.com/mr86ddtu

"Our Approach." Arbor Rising. https://arborrising.org/approach

Partnership for Working Families. 2013. Construction Careers Handbook: How to Build Coalitions and Win Agreements That Create Career Pathways for Low Income People and Lift Up Construction Industry Jobs. https://tinyurl.com/4wfurnfy

"Powering Climate & Infrastructure Careers for All." 2025. The Families and Workers Fund (blog). 2025. https://tinyurl.com/tjttdaj4

"The State of American Jobs." 2016. Pew Research Center. October 6. https://tinyurl.com/yhf25erk

"Washington's Cap-and-Invest Program." 2025. State of Washington Department of Ecology (blog). https://tinyurl.com/bdzbhzac

# About the Contributors

## EDITORS

**Richard A. Benton** is the director of the Climate Jobs Institute and an associate professor in the School of Labor and Employment Relations at the University of Illinois Urbana-Champaign. His research interests include corporate governance, social networks, economic sociology, and social stratification. In general, he focuses on the role of social networks in individual and organizational outcomes as well as the dynamics of network change. Mainly, his research stream examines the consequences of network fracturing among U.S. corporate leaders. Much of his research applies advanced models for network dynamics.

**Lara Skinner** is executive director of Climate Jobs Institute at Cornell University. Her research, writing, and labor education work focuses on the intersection of job creation, economic development, and climate protection. She began her career in labor working with Oregon's Farmworkers Union (Pineros y Campesinos Unidos del Noroeste), the University of Oregon Labor Education and Research Center, and as an active member of the Graduate Teaching Fellows Federation, Local 3544. Skinner has worked for unions doing campaign research and policy development since 1999.

## CONTRIBUTORS

**J. Mijin Cha** researches and writes about climate and environmental justice—how to transition to a carbon-free economy in a way that protects workers and communities, labor–climate coalitions, and the relationship between inequality and the climate crisis. Currently, she is an assistant professor in the environmental studies department at the University of California, Santa Cruz, affiliate faculty in the legal studies department, and on the Faculty Advisory Board at the UCSC Center for Labor and Community. She is also a Fellow at Cornell University's Climate Jobs Institute. Her community-engaged service includes the Board of Greenpeace USA Fund and Emeritus Board at the Center on Race, Poverty, and the Environment. Her book, *A Just Transition for All: Workers and Communities for a Carbon-Free Future*, was published by MIT Press in 2024. Cha's previous work experience includes associate professor at Occidental College, adjunct professor at Fordham University School of Law, and positions with various policy and think-tank organizations.

**Patrick Crowley** is president of the Rhode Island AFL-CIO. He is a union organizer with nearly 30 years of service to the labor movement. During that time he has worked for the Teamsters, SEIU, and NEA Rhode Island. Named one of the most influential Rhode Islanders by *The Providence Business News*, Crowley holds master's degrees in labor studies from the University of Massachusetts Amherst and history from the

University of Rhode Island. In 2021, he helped found Climate Jobs Rhode Island, a broad and growing coalition of labor, environmental, and community partners committed to a just transition to an equitable, pro-worker, pro-climate green economy. He also serves on the boards of several organizations, including the Institute for Labor Studies and Research, the United Way of Rhode Island's Community Advisory Board, Delta Dental of Rhode Island, the Rhode Island Commerce Corporation, the Rhode Island Public Transit Authority, and the Museum of Work and Culture Foundation. He contributed a chapter in the book *Power Lines: Building a Labor–Climate Justice Movement*.

**Zach Cunningham** is a faculty member and the assistant director of Labor Education at the Cornell ILR Climate Jobs Institute. In this role, Cunningham works with labor leaders and members, legislative partners, and state-based climate jobs coalitions to develop educational programs and resources that deepen their engagement and leadership in the fight to tackle climate change through high-quality, union job creation. Before joining the Climate Jobs Institute, Cunningham was a high school teacher and worked as a labor educator with the Civil Service Employees Association, Local 1000 AFSCME. He holds a bachelor's degree from Indiana University and master of industrial and labor relations from Cornell's ILR School, and he currently serves as a vice president for the United Association for Labor Education.

**Brendan Davidson** is a Ph.D. candidate in the Department of Political Science at Colorado State University. Trained in environmental politics from both global and comparative perspectives, he specializes in environmental labor studies, international political economy, and the political ecology of labor in the energy transition. His current research compares various eco-social visions of the transition with the actual practices of work and production associated with low-carbon energy technologies, especially wind energy, doing so primarily in the western United States. Davidson's existing research appears in the journal *Energy Research and Social Science*, with forthcoming research based on his fieldwork under review with journals dedicated to critical political economy and labor process theory. Related to his research, Davidson is a member of the Trade Unions and Labour Environmentalism Network that explores how trade unions can advance working-class environmentalism and tackle today's ecological challenges through collective action. In addition to his work as a scholar, Davidson works for the Center for New Energy Economy, also affiliated with Colorado State University. Prior to pursuing a Ph.D., he worked for several organizations that inform his research interests, including the National Renewable Energy Laboratory, National Park Service, and Colorado's Human Trafficking Council housed within the state's Office for Victims Programs.

**Jessie HF Hammerling** is the co-director of the Green Economy Program at the UC Berkeley Labor Center. She works in collaboration with government, industry, unions, and community stakeholders to understand the impacts of the energy transition and to develop strategies for decarbonizing the economy that generate quality jobs and equitable outcomes. In prior years, Hammerling worked with the

Labor Center's Technology and Work Program, where she studied worker-driven responses to managing technological change. She has a Ph.D. in geography from UC Davis and a master of public policy from the University of Wisconsin–Madison.

**Avalon Hoek Spaans** is the assistant director of research at Cornell University's Climate Jobs Institute at the Cornell School of Industrial and Labor Relations and Cornell's primary investigator on the studies discussed in her chapter. She is credited with developing the New York survey instrument, which was adapted for use in Texas, and for the initial design of both studies. Hoek Spaans specializes in research and policy development at the intersection of equity, workforce, and climate change mitigation and adaptation.

**Jillian Morley** is a research support specialist with Cornell University's Center for Racial Justice and Equitable Futures. She supported the data analysis and interview protocol design for each of the studies discussed in her chapter. Morley's research focus is on social policy mechanisms that exacerbate race and gender inequity in the labor market, healthcare, and housing.

**Hunter Moskowitz** is a doctoral candidate in world history at Northeastern with a B.S. in industrial and labor relations from Cornell University. His dissertation focuses on the movement of labor and technology in the early 19th century textile industry in mill communities such as Lowell, Massachusetts; Monterrey, Nuevo León; and Concord, North Carolina, and how this shaped global understandings of race. He also works on researching climate and labor policy and just transitions at the University of California Santa Cruz.

**Virginia Parks** is a full professor in the Department of Urban Planning and Public Policy at the University of California, Irvine, and the faculty director of the UC Irvine Labor Center. Her research and teaching expertise include labor and employment, urban politics and policy, economic inequality, and local economic development. Her current research focuses on workers and regions impacted by the energy transition and decarbonization efforts. Parks has published research on a range of labor market issues including union training programs, the racial wage gap, immigrant employment, and workplace diversity. She has a Ph.D. in geography from UC Los Angeles and a Master of Urban and Regional Planning from UC Los Angeles.

**Robert Pollin** is Distinguished University Professor of Economics and co-director of the Political Economy Research Institute at the University of Massachusetts Amherst. His books include *The Living Wage: Building a Fair Economy* (co-authored 1998), *Contours of Descent: U.S. Economic Fractures and the Landscape of Global Austerity* (2003), *An Employment-Targeted Economic Program for South Africa* (co-authored 2007), *A Measure of Fairness: The Economics of Living Wages and Minimum Wages in the United States* (co-authored 2008), *Back to Full Employment* (2012), *Green Growth* (2014), *Global Green Growth* (2015), *Greening the Global Economy* (2015), and *Climate Crisis and the Global*

*Green New Deal: The Political Economy of Saving the Planet* (co-authored 2020). He has worked as a consultant for the U.S. Department of Energy, the International Labour Organization, the United Nations Industrial Development Organization, and numerous nongovernmental organizations in several countries and in U.S. states and municipalities on various aspects of building high-employment green economies. He has also directed projects on employment creation and poverty reduction in sub-Saharan Africa for the United Nations Development Programme. Pollin has worked with many U.S. nongovernmental organizations on creating living wage statutes at both the statewide and municipal levels, on financial regulatory policies, and on the economics of single-payer health care in the United States. Between 2011 and 2016, he was a member of the Scientific Advisory Committee of the European Commission project on Financialization, Economy, Society, and Sustainable Development. He was selected by *Foreign Policy* magazine as one of the "100 Leading Global Thinkers for 2013."

**Melissa Shetler** is a senior training and education associate at Cornell University's ILR School Climate Jobs Institute. Shetler began her labor career as a community organizer with the Laborers Eastern Region Organizing Fund, working to build multi-stakeholder coalitions to ensure quality and labor standards in affordable housing. She was the director of organizing and later political director for the Ironworkers Local 46 and served as the executive director of Pathways to Apprenticeship, a direct entry pre-apprenticeship program focused on union career opportunities for residents of public housing and justice-involved individuals. Shetler is a program presenter for Michigan State University's Building Trades Academy, where she teaches strategic organizing. She has an M.A. in adult learning and leadership from Columbia University's Teachers College and a B.A. in interdisciplinary studies from SUNY Empire State College.

**Dimitris Stevis** is a professor in the Department of Political Science and founder and co-director of the Center for Environmental Justice at Colorado State University. He has a longstanding interest in the international political economy of environment and work with particular attention to just transitions. He co-convened the Just Transition Research Collaborative (2018–2022) and is currently a joint coordinator of the Just Transition and Care Network and the Planetary Justice Taskforce and a member of the Transformative Just Transitions Working Group (coordinated by the Global Labour University). Ongoing research focuses on just transition along the lithium life cycle in the United States and the just transition politics of global union organizations. Recent book publications include *Just Transitions: Promise and Contestation* (2023) and *The Palgrave Handbook of Environmental Labour Studies* (2021), co-edited with Nora Räthzel and David Uzzell The latter contributes to the foundation of environmental labor studies as a distinct area of study that recognizes the immanent connection of society and nature. He programmatically pursues engaged research, more recently with the Just Transition Listening Project and the Just Transition and Care Network. He is currently serving as the treasurer of the local American Association of University Professors chapter, a union of educational workers.

**Todd E. Vachon** is assistant professor of labor studies and employment relations at Rutgers University, where he also serves as the director of the Labor Education Action Research Network (LEARN). As the director of LEARN, he oversees the university's labor education programs, including classes, research, and workshops for workers, unions, and other justice-focused organizations. Vachon's research, which has been published in a variety of academic and popular outlets, focuses on inequality, labor, climate change, and justice. His first book, with Tobias (Cornell University Press 2021), puts forth a labor studies perspective on the future of work and workers. His latest book, *Clean Air and Good Jobs: U.S. Labor and the Struggle for Climate Justice* (Temple University Press 2023), focuses on the American labor–climate movement and the struggle for a just transition. His current project, Runaway Inequality and Threats to Democracy, is a guide for popular political economy education programs focused on protecting and extending democracy in order to reduce unjustified inequalities in society. Vachon has a Ph.D. in sociology from the University of Connecticut.

**Jeannette Wicks-Lim** is a research professor at the Political Economy Research Institute at the University of Massachusetts Amherst, where she also earned her Ph.D. in economics. Her areas of research focus include labor economics, with an emphasis on the low-wage labor market; the political economy of racism; the employment impacts of green transition programs in the United States; and the economics of single-payer health care policies. Wicks-Lim is co-author of *A Measure of Fairness: The Economics of Living Wages and Minimum Wages in the United States* (2008). Her forthcoming book (co-authored with Michelle Holder) is *The Political Economy of Racism: The Persistence of Anti-Blackness in the United States*. She regularly serves as an economic policy consultant for nongovernmental organizations on issues of employment and overall conditions for working people in the United States.

**Mike Williams** is a senior fellow at American Progress, where his work focuses on the nexus between creating and retaining high-quality, union jobs, and fighting the climate crisis. He has published pieces on decarbonization pathways for heavy industry, the future of trade policy and its intersection with the climate crisis, and how to ensure clean energy jobs are good, union jobs. Prior to joining American Progress, Williams helped build and lead the BlueGreen Alliance (BGA), serving in many roles over 12 years, most recently as the deputy director. His primary work was to oversee partnership and coalition engagement and advise on and implement the strategic direction of the organization. Williams also helped oversee BGA's policy and advocacy operations. He was point for BGA at the United Nations Framework Convention on Climate Change negotiations from Copenhagen through Paris, and he spearheaded innovative programs—such as BGA's successful Buy Clean effort. Earlier in his career, Williams worked on Capitol Hill. While there, he helped author among the most progressive pieces of climate change legislation at that time. Williams also worked for the National Wildlife Federation, helping launch their environmental education advocacy campaign and assisting with policy and campaigns on energy and water policy. Williams graduated from George Washington University with a master's in public policy, concentrating in environmental policy. He received his bachelor's degree from Boston University, where he studied philosophy and music. He lives in Maine with his wife (Tara), two kids (Marilyn and John), dog, cat, and never enough instruments.

# LERA Executive Board Members 2025–26

**President**
John Budd, University of Minnesota

**President-Elect**
Beverly E. Harrison, Arbitrator/Mediator

**Past President**
Jim Pruitt, Kaiser Permanente

**Secretary-Treasurer**
Andrew Weaver, University of Illinois Urbana-Champaign

**Editor-in-Chief**
Xiangmin (Helen) Liu, Rutgers University

**National Chapter Advisory Council Chair**
William Canak, Middle Tennessee State University (retired)

**Legal Counsel**
Steven B. Rynecki

**Executive Board Members**
Candace Archer, AFL-CIO
Sharon Block, Harvard Law
Andrea Cáceres, SHARE/AFSCME
Maria Figueroa, SUNY Rockefeller Institute of Government
Eliza Forsythe, University of Illinois Urbana-Champaign
Ruben Garcia, University of Nevada, Las Vegas
Janet Gilman, Oregon Employment Relations Boarf
Teresa Ghilarducci, The New School
Steven Greenhouse, Retired, *New York Times*
John Johnson, Southeastern Pennsylvania Transportation Authority
Nelson Lichtenstein, University of California, Santa Barbara
Monique Morrissey, Economic Policy Institute
Hal Ruddick, Alliance of Health Care Unions
Lionel Sims, Jr. Kaiser Permanente
Maite Tapia, Michigan State University
Jonathan Uto, Kaiser Permanente
Marc Weinstein, Florida International University

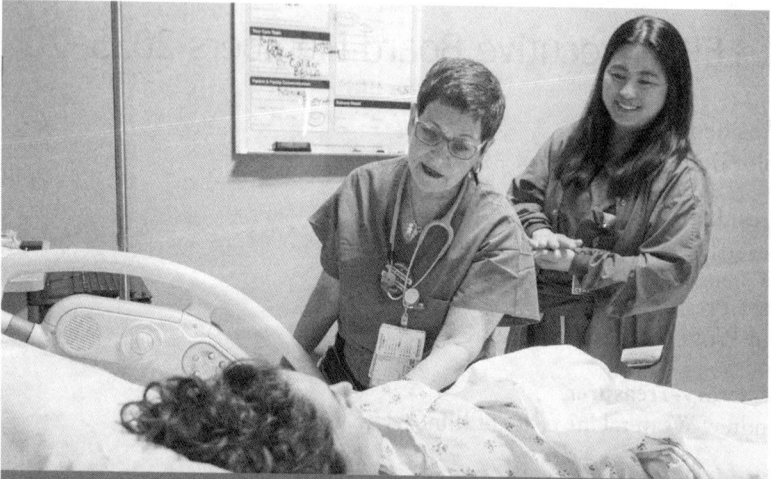

# A *Better* WAY TO →→→→→→→→ WORK

**Kaiser Permanente and the Alliance of Health Care Unions thank all our managers, physicians and employees — including more than 62,000 union-represented workers — for 28 years of partnership.**

Affordable, quality care. It started in California's shipyards and steel mills in World War II; today we continue that tradition across the country. Kaiser Permanente is America's largest non-profit health care delivery organization. Kaiser Permanente and the Alliance of Health Care Unions are proud to work together to ensure Kaiser Permanente is the best place to work and the best place to receive care.

In 2021, we affirmed our Labor Management Partnership in a new four-year agreement, which includes important new provisions committing to joint work on staffing, just culture, and racial justice. Since 2023, joint projects have achieved $230 million in recurring savings promoting the affordability of health care, while enhancing care, service and access. The Alliance-KP partnership includes more than 62,000 members of AFSCME, AFT, HNHP, IBT, ILWU, IUOE, KPNAA, UFCW, UNITE HERE, and USW, in every market where Kaiser Permanente operates.

**KAISER PERMANENTE.**

**ALLIANCE** OF HEALTH CARE UNIONS

# A FOCUS ON RESEARCH

ilr.cornell.edu/research

Cornell University's ILR School is the preeminent educational institution in the world focused on work, labor and employment. ILR-generated research – conducted by the largest number of full-time faculty members focused on work-related topics of any institution in the U.S. – influences public policy, informs organizational strategy and improves professional practice.

ILR faculty and students have produced research on a diverse range of topics, including sexual harassment, remote work, fostering creativity in the workplace, migrant worker rights, the pitfalls of AI and the dangers facing workers in the global apparel and fishing industries.

We are also home to the Catherwood Library, The Kheel Center for Labor-Management Documentation & Archives, and over 15 centers and institutes focused on bridging the gap between research and practice.

**ILR School**

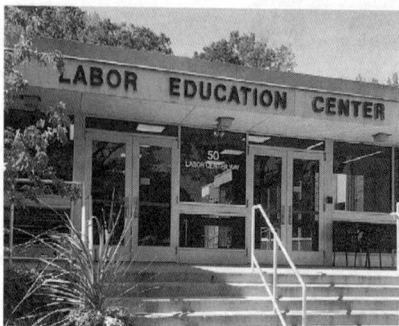

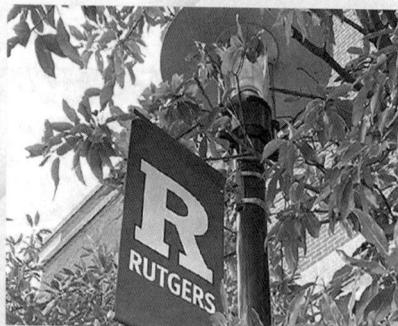

# Leading the Way with LERA

Rutgers School of Management and Labor Relations (SMLR) is proud to partner with LERA through an established tradition of leadership and service to the association across executive, editorial and committee roles. Through our academic programs, research initiatives, and outreach programs, Rutgers SMLR is a leading source of expertise on the world of work, building effective and sustainable organizations, and the changing employment relationship.

Rutgers SMLR is pleased to continue our collaboration with LERA to offer a unique lens to explore the evolution and future of work on a global scale.

**R** | **RUTGERS–NEW BRUNSWICK**
**School of Management and Labor Relations**

smlr.rutgers.edu

School of Labor and Employment Relations

# Climate Jobs Institute

## Our mission

The Climate Jobs Institute at the University of Illinois Urbana-Champaign informs Illinois' clean energy transition through research that foregrounds workers and their communities. We guide state climate policy to reduce emissions and promote high-quality job creation.

## What we research

Clean energy policy

Quality job creation

Workforce & economic development

Labor impacts

Climate justice

## Stay informed

ler.illinois.edu/climate-jobs-institute

# CALL FOR PROPOSALS FOR 2027 RESEARCH VOLUME

The LERA Editorial Committee, chaired by Xiangmin (Helen) Liu, invites you to submit a research book proposal for the LERA 2027 Research Volume.

Proposals will be accepted through **March 15, 2026,** and you will find complete details and a submission form here: https://lera.memberclicks.net/RV-proposal.

The LERA Editorial Committee welcomes submissions on all topics related to labor and employment relations, including but not limited to the following areas:
- AI, Employment Relations, and Worker Rights
- Demographic Diversity, Labor Markets, and Employer Strategies
- Emerging Models of Labor Unionism
- Employer Resistance to Unionization Efforts
- Labor Unions, Political Engagement, and Public Policy
- Innovative Approaches to Dispute Resolution
- Worker Ownership, Participation, and Workplace Democracy
- Work-Life Balance and Worker Well-Being

**Recent Volumes**
- Union Organizing and Collective Bargaining at a Critical Moment: Opportunities for Renewal or Continued Decline? (2024)
- The Evolution of Workplace Dispute Resolution: International Perspectives (2023)
- A Racial Reckoning in Industrial Relations: Storytelling as Revolution from Within (2022)
- Valuing Work(ers): Toward a Democratic and Sustainable Future (2021)
- The upcoming 2025 volume is The Climate–Labor Movement: Lessons Learned and the Promise of an Equitable and Diverse Clean Energy Economy and will arrive in your mail boxes soon. The upcoming 2026 volume is under way and tentatively titled Labor in Global Supply Chains.

**Submission Content**
- A rationale and explanation for the volume (1–3 pp. extended abstract).
- An outline of 10–11 chapters, with a brief description of each chapter and related authors.
- Biographical information and links to vitaes for the editor(s) and contributing authors.
- Editors should secure commitments from chapter authors and communicate that LERA's Editorial Committee needs to approve the volume and may suggest changes.